First Years of Human Chromosomes

Dedication

This book is dedicated to the
pioneers of human chromosome research

First Years of Human Chromosomes

The Beginnings of Human Cytogenetics

Peter S Harper

© Scion Publishing Ltd, 2006

First published 2006

All rights reserved. No part of this book may be reproduced or transmitted, in any form or by any means, without permission.

A CIP catalogue record for this book is available from the British Library.

ISBN 1 904842 240

Scion Publishing Limited
Bloxham Mill, Barford Road, Bloxham, Oxfordshire OX15 4FF
www.scionpublishing.com

Important Note from the Publisher

The information contained within this book was obtained by Scion Publishing Limited from sources believed by us to be reliable. However, while every effort has been made to ensure its accuracy, no responsibility for loss or injury whatsoever occasioned to any person acting or refraining from action as a result of information contained herein can be accepted by the authors or publishers.

Typeset by Phoenix Photosetting, Chatham, Kent, UK
Printed by Replika Press Pvt. Ltd

Contents

Preface vii

Acknowledgements ix

Chapter 1
Beginnings: human chromosomes before 1956 1

Chapter 2
46, not 48: the discovery of the human chromosome number, 1956 29

Chapter 3
1959: the chromosome basis of Down's syndrome 55

Chapter 4
The sex chromosomes 77

Chapter 5
The other autosomal trisomies, 1960 97

Chapter 6
Chromosomes and leukaemia 117

Chapter 7
Technology and nomenclature: the next steps, 1960 137

Chapter 8
Later years: the growth of human and clinical cytogenetics 155

Chapter 9
Conclusions 167

Appendices

1. Some general historical publications on human cytogenetics 173

2. List of those interviewed 175

3. The recordings: key to excerpts included on the CD 177

4. Audio CD of interviews with early workers in human cytogenetics (attached to back cover)

Index 179

Preface

This book attempts to tell the story of the early years of work on human chromosomes, as far as is possible through the memories and writings of those who were actually involved. It does not pretend to be a formal history of the subject, nor does it try to undertake a detailed analysis of how the field developed, but I think that the story is itself an important one and that many people working or interested in genetics as it relates to medicine and human biology will find it interesting.

The idea for the book came during the course of a series of interviews with early workers in different areas of human genetics and the realisation that a high proportion of them had played important roles in the first years of human cytogenetics. The inevitability that in the near future some of these people would no longer be living (others were already deceased) and that their stories were in most cases largely untold, made it important to record them in some permanent and accessible form before it was too late. Thus the plan evolved of writing a framework into which a selection of these stories, recorded and written, could be placed.

The period on which this book focuses is that of the key years 1955 to 1960, when a combination of new technologies allowed accurate studies of human chromosomes for the first time. The earlier years up to 1955, and later developments in the 1960s and 1970s, are only summarised briefly, in two chapters (1 and 8), which frame the more detailed part of the book. To record and analyse these later developments fully is of considerable importance, but would have required most of the book to be devoted to them, and any analysis of recent work also needs a person more closely involved in the increasingly sophisticated scientific techniques than I myself have been.

Each of the main chapters (2–7) thus consists of a narrative, into which I have placed as many quotations from the interviews as possible, together with some excerpts from previous writings by workers directly involved. I have tried to give a short life (with biographical sources) and a photograph of the main figures, while at the end of each chapter one or two of the key original papers are reproduced. A list of more general historical sources is given in Appendix 1.

The inclusion of a CD at the end of this book, giving longer (though still brief) excerpts from ten of the most relevant interviews, should allow readers to gain a more vivid idea of the work and its background by hearing those actually involved recount their memories. More detail about the full recordings and transcripts is given in Appendices 2 and 3.

I have tried to write this book in a style that will make it accessible to all those with an interest in the early history of genetics, not just workers in the chromosome field.

I hope also that science historians will find it interesting, even though the book does not attempt to be an historical analysis; in particular I hope that it will encourage preservation of the primary records of the field, without which future historians may find it difficult or even impossible to undertake the detailed study of this important chapter in human biology and medicine.

Peter Harper
Cardiff, December 2005

Acknowledgements

MANY PEOPLE HAVE CONTRIBUTED TO the creation of this book, but I alone am responsible for any errors or misinterpretations that it may contain. My first thanks must be given to all those who agreed to be interviewed, since it is their memories that have provided both the impetus for writing the book and its key foundations. Many people also most generously gave access to correspondence and loaned photographs, often unpublished and at times entirely unexpected. Their welcome and trust was most touching and has contributed greatly to the pleasure of writing this account of their work. Many also made valuable comments on specific chapters; in the case of John Hamerton this extended to the whole book.

Arnold Munnich in Paris and Ulf Kristoffersson in Lund were of the greatest help during my visits to those two key cities and I am grateful for their generous hospitality.

I have listed all those in the cytogenetics field whom I interviewed in Appendix 1, but am conscious that, through focusing for this book on a narrow time period, there are some where I have on this occasion used little or none of the information given by them. I hope that the recorded interview series as a whole may itself prove to be a valuable archive and resource for future workers more qualified than myself. I am grateful also to those who gave information by telephone or mail, and for permission to quote, often extensively, from previous published and unpublished accounts.

The publishers and I have made every possible effort to obtain the relevant permissions for all photographs and original papers reproduced within this book. Acknowledgements of the source of the material and the copyright holder have been made in the text alongside the reproduced material. If we have omitted to acknowledge anyone that we should have done, then we would be pleased to hear from you so that we can correct such omissions in future printings.

I have been greatly helped by Cardiff University Library Services and Media Resources, especially Mari Ann Hilliar, Adrian Shaw and Mark Bankhead, in obtaining literature and producing the edited disc of recordings. A series of archivists and librarians in different centres has also helped greatly in locating and copying material, making me realise how important their role is in ensuring that the history of the field is preserved. June Williams was invaluable in producing interview transcripts, as was Audrey Budding in keeping track of the growing manuscript and illustrations, while my younger colleagues in the Wales Gene Park office have been not only patient but enthusiastic in helping me to increase my IT skills and overcome problems in this area.

Working with the staff of Scion Publishing, who have ensured the rapid and efficient publication of this book, has been a pleasure.

Finally, I have received strong support from numerous workers in the human cytogenetics field, but I owe a special debt to Professor David Harnden for providing detailed and invaluable criticism on a series of drafts of the entire manuscript, and for his continual encouragement that this was a worthwhile endeavour.

CHAPTER 1

Beginnings: human chromosomes before 1956

THE FOCUS OF THIS BOOK IS ON THE remarkable flowering of work in human cytogenetics beginning around 1955–6, but this means that a century of earlier studies has to be covered in a single brief chapter. This does an injustice to the workers involved, struggling for the most part with hopelessly inadequate material and techniques, and often painfully aware of their own limitations. The most that I can do here is to point out the steps and advances that occurred, and provide some references which will allow readers to follow the story in more detail. The pioneer American cytogeneticist TC Hsu, in his highly readable book *Human and Mammalian Cytogenetics. An Historical Perspective* (1979),[1] salutes what he calls 'the Knights of the Dark Age', many of whom he knew personally, for their efforts. I should like to do the same, even though I never met them.

Early concepts of inheritance

It is hardly surprising that the relevance of chromosomes to inheritance should not have been appreciated until ideas had developed regarding the actual mechanisms of inheritance. The two key fundamentals – a means of ensuring that characteristics of a species are passed on unaltered from generation to generation, yet at the same time allowing variation and evolution to occur – had been recognised in a general way for centuries. Yet, even by the mid-19th century, naturalists, practical plant and animal breeders and more general thinkers had no clear ideas of what these mechanisms were.

More progress during this period was being made with the second of the two questions – how do species change and evolve? – but, as Darwin was to find to his cost, the lack of an acceptable mechanism of inheritance to explain the first question – how do organisms remain unchanged through generations? – had serious consequences for understanding evolution and providing a mechanism for natural selection to work on.

The answer came, of course, in 1865, with Mendel's work[2] showing that a particulate basis for inheritance, with the underlying genetic factors remaining distinct and not blending in the crossed generation, could be demonstrated experimentally, and that this could also be given a simple mathematical basis. However, Mendel's results lay unrecognised until 1900, so Darwin, Galton, and other workers had to grope, with varying degrees of success, towards an acceptable inheritance mechanism.

Darwin's progressive reversion, in his 'pangenesis' theory, to a Lamarckian position, accepting the inheritance of acquired traits, was a real step backwards; it was forced on him, reluctantly, by the estimate of a relatively short age for the Earth's existence by the physicist Kelvin, later found to be erroneous, and most of his close colleagues were unconvinced. Galton's mathematical basis for inheritance was closer to the truth,[3] but it did not point to any specific underlying physical mechanism. Much more relevant was Weissmann's contribution,[4] for, by clearly identifying the germ line as the relevant basis for inheritance, and by eliminating any role in this for the rest of the body's cells, he focused the thoughts and studies of scientists on the germ cells themselves, in both animals and plants, and on their structure and behaviour in cell division. It was these studies which, as they progressed during the second half of the 19th century, began to reveal the physical basis for inheritance and the vital role of the structures that became known as 'chromosomes'.

Chromosomes as cell structures

Serious studies of human chromosomes began only in the 20th century, but to see them in perspective we must look earlier to the 19th-century studies of cell structures, especially the nucleus and its changes in cell division. Plants, insects, worms, larval amphibians and chick embryos were among the organisms used for these studies (Figure 1-1), which formed the foundations for what later became cell biology and are well described in detail in Henry Harris's

TABLE 1-1
EARLY LANDMARKS IN CYTOGENETICS

1840s	Division of cell nucleus in animal and plant cells studied (Remak, Hofmeister)
1850s	First detailed microscopic studies of chromosomes
1887	Constancy of chromosomes through cell generations (Boveri)
1902	Role of chromosomes recognised in Mendelian heredity (Sutton, Boveri)
1912	First 'accurate' analysis of human meiotic chromosomes (Winiwarter)
1914	Chromosome basis of cancer proposed (monograph of Boveri)
1923	Definitive confirmation of human Y chromosome (Painter)
1949	Discovery of sex chromatin (Barr and Bertram)
1956	First publication of correct human chromosome number (Tjio and Levan)

Fig. 1-1 Early images of non-human chromosomes. (a) Mitosis in the salamander larva (from Flemming, 1879),[10] (b) chromosomes of the worm *Ascaris* (from van Beneden, 1883).[15]

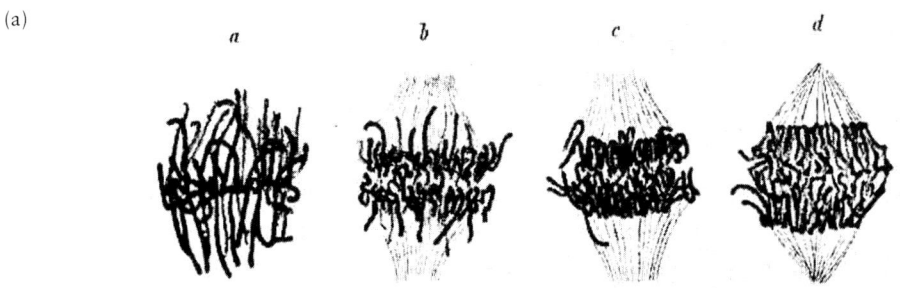

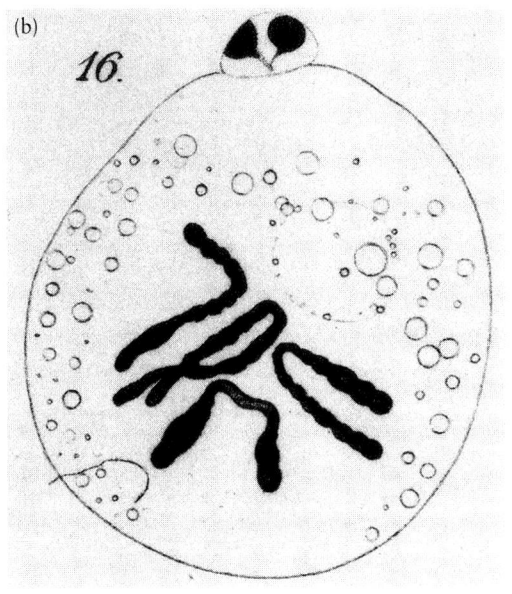

thoughtful and detailed history of the subject, *The Cells of the Body*,[5] which includes many portraits of the workers involved. Table 1-1 summarises some of the main steps connecting this early work with human cytogenetics.

By the 1850s the work in Germany of both Remak (1852),[6] and Virchow (1855)[7] had firmly established the fundamental principle that all cells arise from other cells, '*omnis cellula e cellula*'. The importance of division of the cell nucleus in both plant and animal cells was also recognised by this time, with Remak's 1841 studies[8] on chick embryos and Hofmeister's work (1848)[9] on plant cells.

The term *mitosis* was not used until 1879, by Flemming,[10] but the main steps in the process had already been fully described over the previous 20 years by Hofmeister,[11] Balbiani,[12] and Schleicher,[13] and also the behaviour of the 'rod-like bodies' (later, in 1888, termed 'chromosomes' by Waldeyer)[14] as part of this.

The process now recognised as *meiosis* was documented by studies on the maturation and fertilisation of the egg of the roundworm *Ascaris*. With only four easily distinguishable chromosomes (just two in another variant) and a transparent envelope, van Beneden in 1883[15] was able to distinguish the male and female contributions and show that they remained distinct.

Four years later Boveri[16] not only confirmed this work but developed the concept of chromosomes being specific both in their morphology and function, while also being constant and continuous through cell divisions and generations. These two workers have an additional importance in the present context, for they were the teachers of those scientists who made the first detailed studies of human chromosomes; as will be seen below, van Beneden was the teacher of Winiwarter, while Boveri provided the influence for Sutton and Painter, among others.

Thus, before the end of the nineteenth century, there was strong evidence for the chromosomes being closely connected with the still undetermined process of heredity. The overall state of knowledge at this time is summed up by the first (1896) edition of EB Wilson's classic book, *The Cell in Development and Inheritance*.[17]

Chromosomes and heredity

The rediscovery of Mendel's work in 1900 gave for the first time a clear theoretical basis for the inheritance of characteristics, one which could also be demonstrated by breeding experiments. Despite the fact that some of Mendel's strongest supporters, notably William Bateson, remained sceptical or even hostile to the chromosomal basis of inheritance, the microscopic studies of chromosomes and cell division described above provided an already existing and well documented structural foundation on

THEODOR BOVERI (1862–1915)

Born in Bamberg, Germany, Boveri worked first in Munich and later became Professor of Zoology in Würzburg. He was offered the directorship of the new Kaiser Wilhelm Institute of Biology in Berlin in 1910, and was responsible for its detailed planning, but decided against moving from Würzburg on account of his declining health. Boveri made a series of key contributions to early cytogenetics, including the recognition of the constancy of chromosomes through cell generations, their individual specificity and their role in transmission of hereditary characters, while his 1914 monograph, *On the Problem of the Origin of Malignant Tumours*[24] (later translated by his American born wife and co-worker, Marcella), laid the foundations for understanding the chromosome basis of cancer (see Chapter 6). Numerous investigators from other countries who worked with him included Sutton and Painter. Several biographical memoirs of Boveri have been written in German, but a valuable account of his life and ideas in English is provided by Wolf.[56] Image courtesy of Professor U Wolf and Wiley & Sons, Inc., New York. The photograph was taken in 1909 at the Darwin Centennial meeting, Cambridge.

WALTER SUTTON (1877–1916)

Born to a farming family in Kansas, USA, Sutton became a graduate student of CE McClung at Kansas University, where his observations on chromosome pairing were made, his Master's thesis being on spermatogenesis in the grasshopper *Brachystola magna*, which he had discovered to have exceptionally large chromosomes. His key papers in 1902 and 1903[19,20] establishing the chromosome theory of heredity were written after he had moved to Columbia University, New York, to work with EB Wilson. Sutton left genetics without completing his PhD to enter medical school, always his main aim; he became a distinguished surgeon, specialising in orthopaedic and plastic surgery, but sadly died young from acute appendicitis. Biographical articles have been written by Drs Victor McKusick[57] and James Crow,[58] both themselves distinguished human geneticists. (I am most grateful to the University of Kansas Archive for providing this photograph.)

which the new ideas of genetic transmission could rest.

Both Boveri[18] and Sutton,[19] in 1902, explicitly linked chromosome division in mitosis with Mendel's principles, sex determination being the clearest example of the process, at least in insects. By 1910, the experimental work on *Drosophila melanogaster* by Morgan and his colleagues[21] had begun; the suitability of *Drosophila* for cytogenetic studies (and later the fortunate recognition of the giant salivary gland *Drosophila* chromosomes[22]) allowed detailed correlation of breeding studies and cytogenetic observations. This meant, importantly, that cytogenetics could become an integral part of 'classical genetics' from the very beginning, rather than remain the exclusive province of cytologists, as noted by Carlson in his lucid history of classical genetics, *Mendel's Legacy*.[23]

Early studies of human chromosomes

For the early workers trying to analyse chromosome structure and the processes of cell division, the simpler, rapidly dividing cells of insects and amphibian larvae, with few but large chromosomes, were the obvious materials to use; plant cells, too, proved valuable subjects for chromosome studies, especially for the later experimental studies using such species as *Vicia faba* and *Allium*. It is thus not surprising that early cytogenetics was associated mainly with basic zoology and botany, nor that many later workers on human chromosomes had their basic training in these disciplines. Not just human, but mammalian chromosomes generally must have seemed most unpromising in all respects, especially on account of their small size and large numbers. However, natural curiosity about all things human, especially in such a fundamental matter as heredity, encouraged some people to make attempts, at least, to analyse human chromosomes. The prescient hypothesis of Boveri that cancer might have an underlying genetic basis[24] was also a practical stimulus, so that a number of reports had already appeared by the time that the first technologically adequate analyses were made. Most of these found a human diploid chromosome number of 24 or less, reflecting the almost complete inadequacy of techniques then in use for mammalian cells. Extensive lists of these reports appear in the later studies of Painter[25] and in the book of Makino.[26]

The first worker to produce results of high quality, standing out from all contemporaries, was Winiwarter (often referred to as von Winiwarter or de Winiwarter), working in Liège, Belgium, where he had trained with the pioneer cytologist van Beneden, mentioned earlier. In a series of reports, the first and principal one in 1912,[27] Winiwarter analysed sectioned

HANS WINIWARTER (1875–1949)

Born in Vienna, but coming to Liège, Belgium, aged 3, where his father was Professor of Surgery, Winiwarter (also known as de or von Winiwarter) became Professor of Embryology at the University of Liège. Winiwarter's studies gave the first (1912) close-to-accurate estimate of the human chromosome number, but his insistence that there was no human Y chromosome proved to be incorrect. His deserved reputation as the founder of human cytogenetic studies is based both on his meticulous technique and on his recognition of the limitations of the methods of his time (see text). Winiwarter was a devotee of Japanese literature and art, besides collaborating with the Japanese scientist Oguma; he was also a musical composer, under a pseudonym. Several interesting appreciations of his life and work have been written by later workers in Liège,[59,60] including a detailed reassessment of his cytogenetic studies. (Photograph from Leplat.[60] I am grateful to Professor Lucien Koulischer, University of Liège, for providing details on Winiwarter.)

human testicular material, obtained at surgery, for both the haploid and diploid number, consistently finding 24 as the haploid number in spermatocytes and 47 in diploid spermatogonia, an estimate twice as great as most of the earlier studies.

Winiwarter concentrated his work on the analysis of meiotic chromosomes in spermatocytes and he makes the point, easy to overlook today, that these gave a much clearer basis at that time for identifying and counting chromosomes than did the diploid chromosome set of spermatogonial cells. Thus, as with the later work of Painter, estimates of the human diploid chromosome number were largely made and interpreted in relation to the reference point of the more definitive haploid number, which Winiwarter repeatedly estimated as 24.

Winiwarter, and his Japanese collaborator Oguma,[28] consistently observed an XO pairing mechanism for sex determination in human male meiotic cells, with no evidence of a Y chromosome; they vigorously maintained this view despite the observation of a Y chromosome by others, and they suggested that there would be a sex difference in the diploid number, males (the only sex actually studied directly) having 47 chromosomes and females 48. Winiwarter's papers, especially his 1930 review with Oguma,[29] are spirited, even polemic in style, and he does not hesitate to criticise other workers for poor techniques and misinterpretations. However, he did take the important step of exchanging material with those with whom he disagreed, notably Evans in America, who had proposed a 48 diploid count in both sexes using epithelial cells.[30] Winiwarter and, in a separate analysis but reported in the same paper, Oguma[29] restudied the material from which Evans had derived his 48 counts and found that their own estimates varied considerably; thus, on one preparation the counts were 48 (Evans), 52 (Oguma) and 55 (Winiwarter).

Winiwarter's conclusion in this paper is perhaps his most important contribution to the subject, for it explicitly recognises the limitations of the material that he and the others were attempting to interpret.

'If three cytologists, experienced in the analysis of chromosomes, end up with such discordant estimates, it means that *the images do not possess the type of evidence that allows only a single answer to be imposed.*'
(Winiwarter and Oguma, 1930; author's translation, italics present in original.)

Thus, Winiwarter never pretended that his estimate of a 47 diploid human chromosome number was precise or definitive, but his work had conclusively shown that the number was of this order, far higher than the estimates of around 24 for the diploid number found by most of the earlier workers. It was to be another 40 years from his original study before the techniques would allow an unequivocal count, and that Winiwarter's early estimate proved to be so close to reality is a remarkable achievement.

THEOPHILUS SCHICKEL PAINTER (1889–1969)

Professor of Zoology and later President of the University of Texas, Austin, USA, Painter was born in Virginia and had studied with Boveri in Würzburg between 1913 and 1914. His best known contribution was the recognition of the significance of the giant salivary gland chromosomes in *Drosophila*, but his work with HJ Muller was of the greatest importance in showing the chromosomal basis of mutations and the physical basis for linkage maps. Painter's observations on human chromosomes agreed closely with those of Winiwarter, but he showed definitively that there was a human Y chromosome. An appreciation of Painter has been written by Bentley Glass, himself a student of Painter; a man with strong outdoor interests, it is clear that Painter inspired the greatest respect from his Austin colleagues (who included TC Hsu), both for his research and as University President. He returned to research after stepping down from his administrative role and kept in touch with new developments to the end of his life. It should also be recorded, though, as noted by Hamerton in a recent biographical article,[61] that Painter was President at the time when University of Texas was racially segregated, rejecting an application from an African-American student, a decision overturned by the US supreme court. (Photograph reproduced courtesy of University of Texas, Austin, Archive.)

Winiwarter's long collaboration with Oguma, who spent 10 years in Liège, deserves a note in being one of the earliest European–Japanese biological links, as well as for its lasting influence on Japanese cytogenetics, since Oguma developed a flourishing research unit in Sapporo on his return, being succeeded there by the equally distinguished Sajiro Makino.

For a number of years Winiwarter was the only worker to find such a high chromosome number as 47 in human cells, but in 1921 he received support from the studies of Theophilus Painter, which were to have a profound influence on human cytogenetics for the next 35 years.

Painter was an experienced cytogeneticist who had already made his mark with his work on sex determination in marsupials and who would later become famous for his recognition of *Drosophila* salivary gland giant chromosomes. Like Winiwarter, he used fresh human testicular material, but rather than operative samples he used testes removed from inmates at the Texas 'state insane asylum' on the grounds of 'excessive self abuse coupled with certain phases of insanity'. The ethics of this procedure would be considered dubious now, but seem not to have been thought unacceptable at the time. Immediate sectioning and fixing of the tissue (Painter was in the operating room himself and noted that one patient went to sleep during the procedure) was a major factor in the high quality of the preparations. TC Hsu[1] and Bentley Glass,[31] themselves both at University of Texas, give full background accounts of this work.

Painter's initial 1921 report[32] was brief and mainly concerned the presence of a Y chromosome during meiosis in both the opossum and human. His statement on the human diploid number was cautious, and, as it later proved, completely correct.

> *It will be of general interest to biologists to know that the diploid number of chromosomes for man is very close to the number (47) given by Winiwarter. In my own material the counts range from 45–48 apparent chromosomes, although in the clearest equatorial plates so far studied only 46 chromosomes have been found. Before a final conclusion is made on the exact number it is desired to make a careful study of a large number of division plates. There can be absolutely no question, however, but that the diploid number of chromosomes for both the white man and the Negro falls between 45 and 48.*

By the time of his full report in 1923,[25] however, Painter's views had become more definite. Much of this detailed and fully illustrated paper again concerned meiosis; he found a haploid chromosome count of 24, the same as Winiwarter, and he confirmed the presence of an XY bivalent – unlike Winiwarter, who had repeatedly found an XO human male karyotype, though Painter considered that this resulted from a different interpretation of the bivalent structure, rather than different observations. Given Painter's certainty regarding his sex chromosome findings and his agreement with Winiwarter on the haploid count of 24, it is not surprising, as noted by Malcom Kottler in his historical reassessment (to be discussed later) and as Painter himself states in his paper, that a diploid count of 48 would be expected, with Winiwarter's own count of 47 explained by his not having recognised (or misinterpreted) the small Y chromosome.

It does not appear that Painter and Winiwarter ever exchanged material, as Winiwarter had with Evans, and as Painter also offered in his published paper. Perhaps Winiwarter was apprehensive that Painter might find a Y chromosome in the material he could have sent him.

Painter does not explain how his earlier 1921 comment favouring 46 chromosomes evolved to that in 1923, when his 'best' cases (those chosen for illustration) all showed a diploid number of 48. It seems likely, as recognised by Winiwarter and as discussed later in relation to the work of Tjio and Levan, that the preparations simply could not allow such a specific conclusion, that his earlier statement of the number being 'between 45 and 48' was indeed the most appropriate one, and that the specific number of 48 was largely derived from what was to be expected from the apparent haploid number of 24. Neither Painter nor Winiwarter were to know that this haploid number was wrong, and that there are in fact 22 human autosomes, not 23. However, the agreement on the 24 haploid number by both these major and authoritative studies, along with Painter's conclusion of 48 for the diploid number, must have formed a benchmark for later workers when trying to interpret equally difficult material.

The meticulous work of both Winiwarter and Painter, even though disagreeing over the mechanism of sex determination and whether a Y chromosome was present in man, agreed that there was a haploid chromosome number of 24 and that there were 23 pairs of autosomes. By the mid-1920s, therefore, there must have seemed little point in further detailed studies on normal human material, especially since this remained largely limited to testis for preparations of adequate quality.

There were such studies undertaken, however, and almost without exception both testicular and somatic cell counts confirmed Painter's conclusion of 48 as the diploid number, with a haploid number of 24 in spermatocytes. Makino's 1956 review[33] and later book list these in detail and Table 1-2 summarises the principal ones. In none is the conclusion reached that the diploid number might be 46 or that the haploid number could be 23, something that today appears remarkable, as discussed in the next chapter. As already noted, what was wrong was not the work itself, but the

TABLE 1-2
EARLY STUDIES OF THE HUMAN CHROMOSOME NUMBER, 1912–1955

Date	Author	N	2N	Tissue
1912	Winiwarter	24	47	Testis
1922	Oguma and Kihara	24	47	Testis
1923	Painter	24	48	Testis
1927	Evans and Swezy	24	48	Testis
1928	Kemp		48	Tissue culture
1930	Oguma	24	47	Testis
1930	Swezy and Evans		48	Ovary
1935	Chrustschoff and Berlin		52	Blood
1936	Andres and Navashin		17–48	Testis
1945	Slizynski		48	Bone marrow
1952	Hsu		48	Tissue culture
1953	Sachs		48	Endometrium
1955	Manna		48	Uterine cervix

(Based on Makino, 1975. Only studies of normal tissue included.)

attempt to extract a more precise conclusion from it than was justified by the actual material. For a truly accurate definition of the number and properties of human chromosomes, new technologies were needed and they had to be applied together.

Kottler, in his valuable review on the topic,[34] one of the very few studies by science historians in this field, blames the intrinsic conservatism of cytologists for the fact that it took another 30 years for this combination of technologies to be used, even though most of them were not actually new. Whether cytologists are, or were, really more reluctant to be innovative than other scientific disciplines is debatable, but since technology proved to be both the limitation to accurate knowledge of the human chromosome number and the key to eventual progress in human cytogenetics, these different elements of technique must now be carefully examined.

Advances in technique

Both Kottler[34] and Hsu[1] defined in their accounts four technological advances that were essential for modern cytogenetic studies and these are listed in Table 1-3. I have added two further factors which seem to me of equal importance: the use of photomicrography rather than camera lucida drawings to give an objective record of chromosome number and morphology; and the availability of human embryonic material, with its intrinsically rapid cell division.

The first two advances came directly

TABLE 1-3
TECHNOLOGICAL FACTORS IN THE STUDY OF HUMAN CHROMOSOMES

Colchicine for arrest of mitosis
'Squash' technique to bring chromosomes into two-dimensional plane
Cell culture techniques, especially monolayers
Hypotonic treatment to spread chromosomes
Photomicrography as objective record
Embryonic tissue with rapid cell growth

from plant cytology, which had made rapid progress, both practical and theoretical, in the understanding of chromosome structure and function under workers such as Darlington and others. A recent biography of Darlington[35] shows how far plant cytogenetics had already progressed before the onset of World War II.

Squashing of cells, first introduced as long ago as 1921 by Belling,[36] essentially converted the three dimensional cell structure into two dimensions and removed the need for fixing and sectioning of the material. One can see how workers on animal tissue, with a long tradition of histology and cutting tissue sections, must have regarded such a crude approach as heretical, yet the problem of fragmentation of chromosomes in a microscopic section and of trying to trace them in a three-dimensional structure was a major limitation to accuracy and reproducibility.

The somewhat empiric, to the outsider almost horticultural, nature of the 'squash' technique has given rise to its own legends; some workers, such as Tjio, were said to be especially adept because of their broad thumbs, usually 'stained with orcein'. Others by contrast, found it difficult for comparable anatomical reasons. TC Hsu, in one of the many anecdotes in his book that make it a pleasure to read, though possibly meeting the disapproval of those more politically correct, provides an example.

Some persons whose thumbs cannot bend upward have trouble doing squashes and Hans Stich is one of them. When Hans was working in my laboratory as a guest investigator, I actually saw him laying all the slides on the floor and stepping on each with his heel, muttering 'I must get my wife down with me the next time I come'. Indeed he did, and I found Kikki to be one of the world's best squashers, in addition to being a beautiful woman.

The use of *colchicine* to arrest mitosis in metaphase, by interfering with formation of the spindle, also came directly from plant cytogenetics in the 1930s; indeed, it is of interest that Albert Levan, one of the first to describe the technique in 1938[37] when based at the Swedish plant breeding centre of Svalof, near Lund, started his career as a plant cytologist, but later turned his research completely towards human cytogenetics, in particular cancer cytogenetics, as also did Joe Hin Tjio; the key role of these two workers is described in Chapter 2. Colchicine greatly increased the number of mitoses available for analysis in a preparation, though the greater degree of condensation of the chromosomes also tended to increase problems of crowding and clumping.

The third technological advance, the use of *cultured cells*, was based on the development of tissue culture methods in the first decades of the 20th century and the story of this is well told in Harris's book, *The Cells of the Body*.[5] In some ways, however, the classical tissue culture approaches, based on solid explants and with the initial emphasis on differentiation and organogenesis, were a hindrance to cytogenetics; much more relevant were the production of monolayers and the use of techniques such as feeder layers to encourage continued cell growth and division, along the lines of those techniques used in microbiology. Although some early human tissue culture studies, such as that of Kemp in Denmark[38] on embryonic tissues, did manage to reach the then 'gold standard' of 48 chromosomes, it was the 'cell cloning' techniques of Puck and colleagues,[39] including TC Hsu, that gave cell cultures which, when combined with the use of colchicine and squashing, were especially suitable for chromosome analysis.

The fourth, and possibly the most important technological advance that made accurate cytogenetic analysis of mammalian cells possible was the use of *hypotonic solutions* to swell the cells and separate the dividing chromosomes. To appreciate just how important a step this was one has only to look at the figures in the early papers and in monographs such as that of Matthey, who brings together all the pre-1950 work on vertebrates in a series of drawings that are beautiful but of limited use. Figure 1-2 shows his human examples. The overwhelming impression given to the modern observer is of a tangled mass of chromosomes, often densely clumped and making it impossible or close to impossible to distinguish individual complete chromosomes from chromosome arms. Putting this limitation together with the problem of chromosomes being in three dimensions prior to squashing leaves one in no doubt that the valiant attempts of the early workers were bound to be inaccurate and that definitive studies on species with large chromosome numbers, such as humans, required some way of unravelling the chromosomal tangle.

Hypotonic solutions proved to be the answer to the problem and the history of their use in cytogenetics deserves a detailed study from science historians because here indeed is a case where one cannot take the accounts of scientists at their face value. TC Hsu is generally credited with their first use to give high quality chromosome preparations and in his book he describes it as an 'accident' from a technician having made up the wrong dilutions, causing a puzzle only worked out by deliberately changing

TC HSU (1917–2003)

Born in Chekiang province, China, Hsu initially studied genetics under CC Tan, but came to America to work at University of Texas, Austin, in 1948. He joined the laboratory of Charles Pomerat to work on cell culture where he discovered the effect of hypotonic saline on spreading chromosomes; subsequently, he focused on studies of cultured cells in cancer. Hsu's technical developments laid many of the foundations for modern cytogenetics and he had a strong influence on the next generation of human cytogeneticists. His 1976 book gives a unique personal perspective on the early years of this field and reflects his lively personality and sense of humour, as does a recent obituary.[63] (Photograph reproduced courtesy of Cytogenetic and Genome Research and S. Karger AG, Basel.)

each variable over the following three months, since no-one would admit to the error! Hsu's inadvertent heroine thus stayed anonymous, though, chivalrous as ever, he admits that:

Fig. 1-2 Early preparations of human chromosomes (plate from Matthey,[62] Figure 28). This selection of early studies shows how the unspread and tangled chromosomes made a conclusive determination of the human chromosome number impossible without interpreting the data beyond what the techniques could allow. Matthey's remarkable book documents and illustrates all known chromosome studies of vertebrates before 1950.

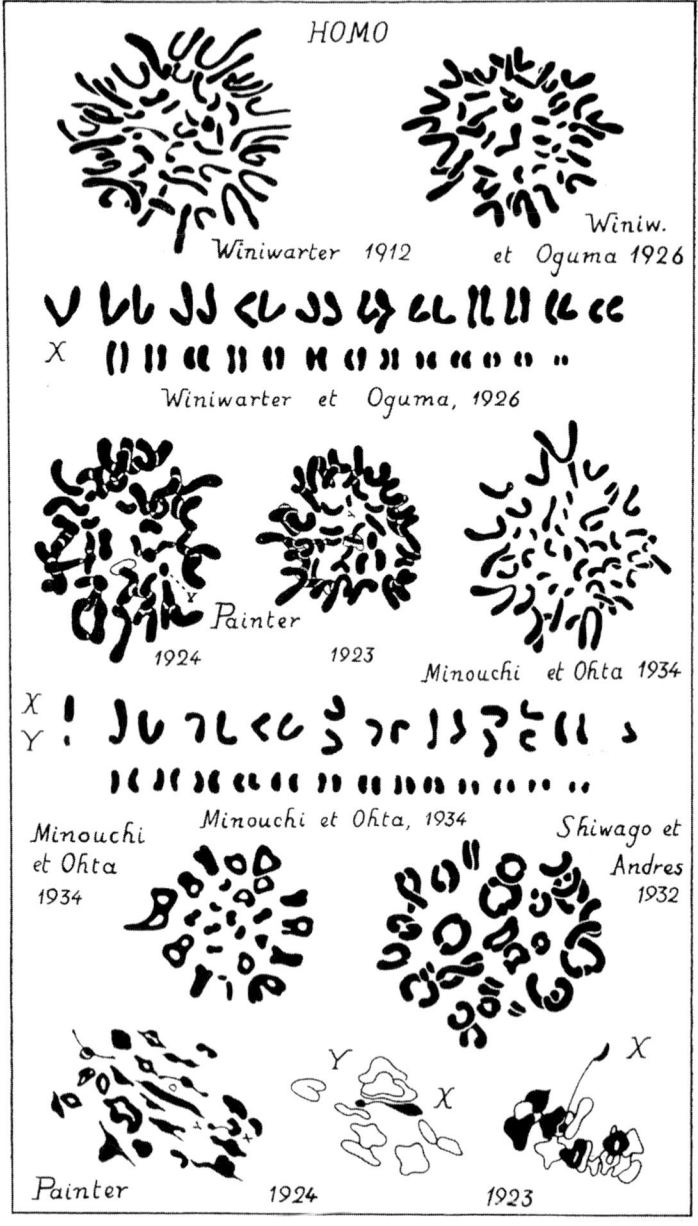

even today, I would love to give a peck on the cheek to that young lady who made such an important contribution to cytogenetics.

Hsu's medical school Dean, Chauncey D Leake, was even moved to compose a poem on the event, in a style reminiscent of the classical poets a century and a half before.

Chance and the Human Chromosomes
In circled nucleus
The twisted rods lie flat
And silhouette their pregnant shapes
For ease in recognition.
A chance mistake
Has pulled away the drapes
Of overlapped confusion on
The forms and faces of
The human chromosomes,
So now they may
Be named, saluted, while
They stay encased
In their old plasmic homes...

Hsu's story is indeed delightful and vivid, but whether it happened quite as accidentally as it appears in his book is something that Harris,[5] in particular, has questioned. Hsu acknowledges that Hughes, working on tissue culture methods at the Strangeways Laboratory in Cambridge, UK, had been systematically investigating the effects of hypotonic conditions, though not specifically on chromosomes, and that he had sent to Hsu a preprint of his 1952 paper.[40] At the same time Makino in Japan

SAJIRO MAKINO (1906–1989)

Makino, along with his teacher Oguma, represents a long and distinguished line of Japanese work in human cytogenetics. Born at Narita, near Tokyo, he studied agriculture at Sapporo University and began human chromosome studies there with Oguma in 1928, later becoming first director of the Chromosome Research Unit and training numerous students. His classic book, *Human Chromosomes*,[26] contains much original material as well as providing a synthesis and detailed sources of early work. Significantly, his obituary notes that he was '*one of the few scientists who refused to carry out the military-backed program*' during World War II. Hsu notes that Makino's deafness resulted from an injury inflicted at this time by a Japanese army officer. (I am most grateful to Professor Michihiro Yoshida, Sapporo, for providing the photograph.)

had observed the spreading effect of hypotonic solutions on chromosomes, again published in 1952;[41] Hsu quotes Makino as later telling him that his discovery was also accidental, resulting from being called away while his slides were in tap water, but this is not mentioned in Makino's own book.[26] An obituary of Makino, however, states that he accidentally left a grasshopper testis in the sink and found the next morning that it contained much clearer and better spread metaphases than before its exposure to tap water. Finally, Russian cytogeneticists had already observed the effect in the 1930s, as part of a truly remarkable programme of cytological research (see below), but their 1936 publication was soon to be overwhelmed along with most of Russian genetics in the debacle that followed shortly afterwards.

Regardless of these details and how 'accidental' the discovery was, there can be no doubt that the 1952 hypotonic solution development, together with the other advances mentioned, allowed TC Hsu to produce human chromosome preparations of previously unrivalled quality. Aided by the additional factor that he had access to human embryonic tissue, he was now in a position to make what should have been the definitive study of human chromosomes. Indeed, he proceeded to undertake just this, though since he was still using a plasma clot culture he was not able to produce as satisfactory a two-dimensional squash preparation as he wished. However, the main factor in his result was not technological but psychological, as will be discussed further in the next chapter – armed with the combination of new techniques, but still encumbered with the weight of accepted tradition, he looked for, and duly found, a human chromosome number of – 48! The true solution to the human chromosome number had to wait for the combination of modern technology and an open mind, and is described in Chapter 2.

The final two factors listed in Table 1-3, the use of *embryonic tissue* and of *photomicrography*, both played important roles in the advances described in the following chapters. *Embryonic tissue*, as we have seen, was used by TC Hsu, but had been employed in earlier studies by Kemp in Denmark in 1924[38] and by the Russian workers, Andres and Jiv, in 1936,[42] who had even been able to analyse chromosomes of the human oocyte using this material.[43] Embryonic tissue was the most satisfactory basis for studying cultured somatic cells prior to the much later (1960) discovery of phytohemagglutinin, owing to its high rate of spontaneous cell division (although, again, as described below and in Chapter 6, the Russians had already managed to stimulate mitosis in peripheral blood cells by 1931!).

Once the use of squashing techniques and cell culture had made it possible to observe chromosomes in a single plane, *photomicrography* was a much less laborious and less subjective way to document chromosome preparations than camera lucida and other drawing techniques. Again, though, some workers, notably Levan, were reluctant to abandon the traditional approaches which involved so much effort as well as skill. When one looks at the drawings involved in works such as Matthey (1949), based on unspread and unsquashed sectioned material, one can sympathise, but there is no doubt that drawing of chromosomes made it much easier to interpret ambiguous material according to what the investigator expected to find.

Russian contributions to early human cytogenetics

A series of remarkable contributions to human cytogenetics, now in danger of being forgotten, was made by Russian workers led by AH Andres, in the Moscow Medical Genetics Institute headed by Solomon Levit. These are mentioned individually in specific chapters, but the research programme as a whole deserves wider recognition as it made a number of key discoveries 20 years or more before they were 'rediscovered' in the West. Notable contributions include:

- use of hypotonic solutions for spreading of chromosomes (Zhivago et al, 1934);[44]
- analysis of chromosomes from cultured peripheral blood, using haemolysed red blood cells as a mitotic stimulator (discussed further in Chapter 7) long before discovery of the similar effects of phytohemagglutinin in 1960 (Chrustchoff et al, 1931;[45] Chrustchoff and Berlin, 1935)[46];
- analysis of cultured embryonic cells (Andres and Jiv, 1936);[42]
- chromosome analysis of human oocytes (Andres and Vogel, 1936);[43]
- detailed morphological analysis of the larger human chromosomes (Andres and Navashin, 1936);[47]
- cytogenetic studies in leukaemia and other cancers (discussed in Chapter 6) (Andres and Shiwago, 1936).[48]

Tragically, the entire research programme had to be discontinued after 1937, when political ideology, orchestrated by the demagogic and largely fraudulent agronomist Lysenko, acting on behalf of Stalin, brought virtually all work on genetics to a halt for more than 30 years (Medvedev, 1969;[49] Soyfer, 1994[50]). The Medical Genetics Institute was closed, its director Levit and others, including the basic cytogeneticist Levitsky, were shot, while remaining workers, if able to continue at all, were forced to work on other topics. The equally important research in the separate institute of Koltsov, involving basic chromosome and molecular studies, was also totally stopped. It seems, though, that none of the human cytogenetics group actually died in this catastrophe (personal communication from Professor Nikolai Bochkov, 2005).

At present, this story has to be pieced together from various translated sources and other works focusing mainly on agriculture and on eugenics,[51] although recent accounts of this period exist in Russian. It is greatly to be hoped that a full account of the work of these outstanding scientists will be given in English.

Lionel Penrose, who was fully aware of the value of this work and of the tragedy of

Russian genetics, aptly sums up the situation in his 1966 review of human chromosomes.[52]

It is natural to infer that, had these laboratories in the USSR not been closed for political reasons, most of the discoveries made in the last nine years on the human karyotype might have appeared some 20 years earlier.

The sex chromatin

The story of human cytogenetics, as told so far, has been one of slow but progressive development, albeit with periods of stagnation and, at the point reached so far, a false illusion that the human chromosome number was known with certainty. Still, the work described has had an orderly sequence, based on clear goals and illustrating the importance and limitations of technology in achieving these goals.

Quite unexpected, therefore, were the results of another line of research, initially unrelated to and unconnected with genetics, and showing how important insights can arise in science if the workers involved are sufficiently alert and curious about the wider aspects of their findings. A remarkable example of this is provided by the discovery of the sex chromatin (Figure 1-3).

In 1949, while human cytogenetics was still awaiting the widespread application of hypotonic spreading of chromosomes and other key techniques, two Canadian neuroscientists, working in London, Ontario, were making microscopic observations on cat neurons in response to electrical stimulation. Recently returned from service in the Canadian Air Force in World War II, Murray Barr was attempting to analyse the neuronal basis of prolonged fatigue, something that had been of much relevance to pilots during the war and he had been joined by Ewart (Mike) Bertram as research assistant (see box). They were intrigued to observe a small body in the nucleus of the neuron, apparently connected to the

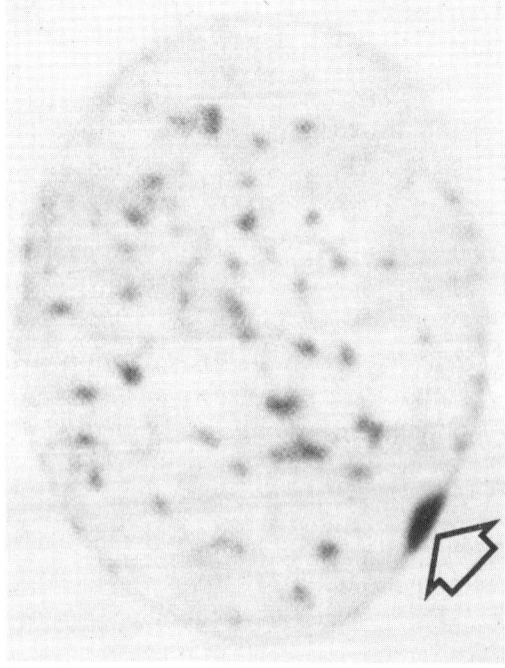

Fig. 1-3 The sex chromatin (from Moore[54]). Discovery of the sex chromatin body in interphase cell nuclei by Barr and Bertram in 1949 allowed important conclusions on human sex determination and sex chromosome abnormalities before full cytogenetic analysis was feasible.

Discoverers of the sex chromatin body, Murray L Barr and Ewart G (Mike) Bertram. Portraits from Moore, 1966, *The Sex Chromatin*;[54] the signatures are original, from a personal copy of the book kindly given by Dr Keith L Moore. (Reproduced courtesy of WB Saunders, Philadelphia.) Moore's valuable book brings together contributions from a wide range of workers on different aspects of the sex chromatin.

MURRAY L BARR (1908–1995)

EWART G (MIKE) BERTRAM (born 1923)

Murray Barr was born in Belmont, Ontario, and trained in Medicine at University of Western Ontario, where he spent his later career as Professor of Anatomy. Starting research in neuroanatomy in 1936, his work was interrupted by World War II, but his decision on return from this to study synaptic fatigue in neurons, was influenced by his wartime air force experience and thus led to the serendipitous 1949 discovery of the sex chromatin body. Despite the extensive international links and collaborations with geneticists over the following decade, Barr essentially remained a neuroanatomist.

Ewart (Mike) Bertram joined Murray Barr as postgraduate student in 1947 after graduating from University of Western Ontario, and recognised that the 'paranucleolar body', which he and Barr were studying in cat neurons, was only present in females. Like Barr, Bertram remained an anatomist throughout his career, having passed on his sex chromatin work to Keith Moore soon after publication of the initial study.

nucleolus, that appeared to move in relation to retrograde stimulation of the nerve trunk. This structure, the nucleolar satellite or paranucleolus, was in fact not new, having originally been described by the great Spanish neuroanatomist Cajal in 1909,[53] but its response to nerve stimulation clearly demanded further investigation, which Barr and Bertram proceeded to undertake between 1947 and 1949.

Thus far the work had no connection with genetics, nor was it intended to, until Bertram found that some of his animals showed no sign of having the structure. On carefully checking his records he found that these were all males – presence of the body was directly related to sex.

Keith Moore, who succeeded Bertram as Barr's co-worker, has carefully documented the history of the discovery in his 1966 book, *The Sex Chromatin*,[54] which brings together all the early work on the topic. Moore's historical chapter apparently went through eight drafts before those concerned agreed on all points, so it can probably be regarded as the definitive story. I was able to interview Mike Bertram and Keith Moore together 55 years later and it can be seen from the following excerpt, and from the longer sequence given in the recording accompanying this book, that the discovery retains its vividness and immediacy for them after more than half a century.

PSH. *Can I come back now to 1948/49 and ask just how did it happen that you noticed this unusual structure?*

MB. *Well, if I hadn't kept very, very careful records probably the discovery never would have been made for maybe a number of years following that. How we actually discovered it was, I was measuring this little chromocentre after prolonged activity and there was depletion of the protein substance in the nerve cells and there was a lot of swelling; the protein, Nissl substance, was all dissolved and this little structure was pushed away. This chromocentre...*

KM. *It was stimulating the hypoglossal wasn't it?*

MB. *Yes, to do all that I stimulated, activated the hypoglossal nerve. The reason for doing that was, in sections you had the normal control side as the left side and stimulating the right side, so under the microscope you had normal cells, supposedly, and abnormal activated cells on the other side. I was measuring the migration of this little structure away from its usual position, and had to go through a whole series of cats because we sacrificed them after so many hours of stimulation, up to 48 hours and then 3 days, 4 days etc; so looking at all the tissue, I had come on to the section of this one cat, drew all the measurements and the next cat the same way and then it came to one...*

KM. *Couldn't find it!*

MB. *Then we started looking up my records because I kept track of the age and the sex and the coloration and all the rest of it. So all these things now we're recording were sex female. So we began looking back at records and going through the series. Every time it appeared in the animals they were female. Then when it didn't appear it was male. We were attributing it to very poor staining and other things that didn't show up. So that's basically how we made the discovery.*

PSH. *Can I ask, at that time what did you think this body actually was? Did you think it was anything genetic?*

MB. *Well at that time no.*

KM. *We thought it was RNA first and then, by Feulgen stain, it showed up as actually DNA.*

PSH. *So you didn't really think you were looking at a chromosome?*

MB. *No. Absolutely not.*

Once their own initial scepticism had been overcome, and the specificity and constancy of the finding confirmed by looking at their considerable body of earlier material, Barr and Bertram published their findings as a short report in the 30th April, 1949 issue of Nature.[55] The paper is given at the end of this chapter and it can be seen that by this time the authors had given careful thought as to what this body might be and had concluded that it was likely to represent the second X chromosome of the female. Today this might seem to be an obvious conclusion, but in 1949 this was far from the case since the X chromosome could not yet be identified directly, except when pairing, and there was even disagreement about the chromosome number and sex determining mechanism in the cat. Matthey, reviewing the different studies in his book,[62] also published in 1949, concluded that most were agreed on a 38 diploid chromosome number and an XY sex determining mechanism, but Winiwarter, in an interesting parallel to his work on human chromosomes already described, insisted on an XO mechanism, with 35 chromosomes in the male cat and 36 in the female.

Soon after this initial work Bertram moved to Buffalo, New York, to undertake a PhD (London, Ontario did not have a PhD programme at this point) and Keith Moore took his place to develop the work, showing that the 'sex chromatin body' (as it soon became known) was present in a wide range of mammals (though not all), including humans, and that it was not a specifically neuronal structure but present in all tissues.

The possibility of a cytological test that would definitively distinguish a chromosomal female from a male had obvious implications for medical diagnostic tests, so a simple non-invasive test became important. Moore's development of the 'buccal smear' test provided this and had an interesting origin.

KM. *We did the skin biopsy first and that made it easier to get specimens, so we got specimens sent to us from all over the world with these different conditions and we would examine them and send them back and then publish papers on that and then I'd take a buccal smear and the story on that is simple.*

MB. *Clever!*

KM. *I was at a meeting and I was showing this, and one lady came up and said 'I work with ducks' and I said 'well I don't know whether ducks have it, but the only way to find out is to take a piece of skin and check it out and see if there's any sex chromatin in it.' 'Oh I couldn't do that with my poor ducks!', she said, 'can't I just scrape the mouth of the duck?' I said 'sure, you would probably get some cells and do it.' Oh boy, that just clicked in my head. I went home and I started scraping; I got one of these metal spatulas; I scraped my own, I scraped my wife and my baby daughter, who is now 50, and they showed up beautifully because you just had to smear it on. Those days we used some kind of an egg fixer...*

MB. *Egg albumen.*

KM. *Egg albumen, then it got that you didn't even have to do that because you smeared it. Stuck it in alcohol and stained with, what was that blue stuff?*

PSH. *Alcian blue?*

KM. *And boy, it would show up beautifully. So once we had that we could smear everything.*

For the first time, clinicians and research workers now had a simple, non-invasive and inexpensive test for a human chromosome that could be used both in the investigation of individual patients and in large-scale studies of particular groups. It would be 10 years until comparable studies could be done by full chromosome analysis, and 20 years until banding techniques allowed the human X to be reliably distinguished from morphologically similar chromosomes. Study of the sex chromatin thus gave an important head-start to the cytogenetic study of disorders of sexual and gonadal differentiation, such as Turner and Klinefelter syndromes, that will be described in Chapter 4. With the knowledge available from sex chromatin analysis, these later chromosome studies could be undertaken with a clear hypothesis as to what result might be expected.

The discovery of the sex chromatin body had taken Barr, Bertram and Moore away from their parent field of neuroanatomy into the unfamiliar and totally different world of genetics, an exciting if unexpected excursion that lasted for over a decade, and with which they will always be associated. It is of interest, though, that all three remained neuroanatomists, returning to work in their own discipline rather than converting to that of genetics, and all three had distinguished careers in Canada as anatomists. Perhaps this underlies the excitement that can still be sensed in talking with Bertram and Moore over 50 years later; one senses that they, with their leader Murray Barr, had visited and explored a previously unknown and uncharted country, but that at the end they needed to come home to familiar territory.

References

1. Hsu TC (1979). *Human and Mammalian Cytogenetics. An Historical Perspective.* New York, Springer.
2. Mendel G (1866). Versuche über Pflanzenhybriden. *Verhandlungen des Naturforschenden Vereines in Brünn.* **4**, 3–47.
3. Galton F (1889). *Natural Inheritance.* London, Macmillan.
4. Weissmann A (1885). The continuity of the germ-plasm as the foundation of a theory of heredity. English translation in: *Essays upon Heredity and Kindred Biological Problems*, Vol. 1. Oxford, Clarendon Press, pp. 167–254.
5. Harris H (1995). *The Cells of the Body. A History of Somatic Cell Genetics.* Cold Spring Harbor, CSHL Press.
6. Remak R (1855). *Untersuchungen über die Entwickelung der Wirbeltiere.* Berlin, G Reiner.
7. Virchow R (1855). Cellular-Pathologie. *Arch. Pathol. Anat.* **8**, 3–39.
8. Remak R (1841a). Über Theilung Rother Blutzellen beim Embryo. *Med. Z. Ver. Heilk. Pr.* **10**, 127.
9. Hofmeister W (1848). Ueber die Entwicklung des Pollens. *Z. Bot.* **6**, 425–434.
10. Flemming W (1879). Beiträge zur Kenntnis der Zelle und ihrer Lebenserscheinungen. *Arch. Mikrosk. Anat.* **16**, 302–436.
11. Hofmeister W (1849). *Die Entstehung des Embryos der Phanerogamen.* Leipzig, Friedrich Hofmeister.
12. Balbiani EG (1876). Sur les phénomènes de la division du noyau cellulaire. *C. R. Acad. Sci. Paris* **83**, 831–834.
13. Schleicher W (1879). Die knorpelzelltheilung. *Arch. Mikrosk. Anat.* **16**, 248–300.
14. Waldeyer W (1888). Ueber Karyokinese und ihre Bezeihungen zu den Befruchtungsvorgängen. *Arch. Mikrosk. Anat.* **32**, 1–122.
15. van Beneden E (1883). Recherches sur la maturation de l'oeuf et la fécondation. *Ascaris megalocephala. Arch. Biol.* **4**, 265–640.
16. Boveri T (1887) Über die Befruchtung der Eier von *Ascaris megalocephala. Sitzungsber. Ges. Morphol. Physiol. München* **3**, 153.
17. Wilson EB (1896). *The Cell in Development and Inheritance.* New York, MacMillan.
18. Boveri T (1902). Über mehrpolige Mitosen als Mittel zur analyse des Zellkerns. *Verh. Phys-Med. Ges. Würzburg* **35**, 67–90.
19. Sutton WS (1902). On the morphology of the chromosome group in *Brachystola magna. Biol. Bull.* **4**, 124–139.
20. Sutton WS (1903). The chromosomes in heredity. *Biol. Bull.* **4**, 231–251.
21. Morgan TH (1910). Chromosomes and heredity. *Am. Nat.* **44**, 449–496.
22. Painter TS (1934). Salivary gland chromosomes and the attack on the gene. *J. Hered.* **25**, 464–476.
23. Carlson EA (2004). *Mendel's Legacy. The Origin of Classical Genetics.* Cold Spring Harbor, CSHL Press.
24. Boveri T (1914). *Zur Frage der Entstehung Maligner Tumoren.* Jena, Fischer. (English translation, *On the Problem of the Origin of Malignant Tumours*, by Marcella Boveri.)
25. Painter TS (1923). Studies in mammalian spermatogenesis. II. The spermatogenesis of man. *J. Exp. Zool.* **37**, 291–335.
26. Makino S (1975). *Human Chromosomes.* Tokyo, Igaku Shoin.
27. Winiwarter H (1912). Études sur la spermatogenèse humaine. *Arch. Biol.* **27**, 91–189.
28. Oguma K and Kihara H (1923). Études des chromosomes chez l'homme. *Arch. Biol.* **33**, 493–514.
29. Winiwarter H and Oguma K (1930). La formule chromosomale humaine. *Arch. Biol.* **40**, 541–553.
30. Evans HM and Swezy O (1929). The chromosomes in man: sex and somatic. *Mem. Univ. Calif.* **9**, 1–64.
31. Glass B (1990). Theophilus Schickel Painter. *Biogr. Mem. Natl. Acad. Sci.* **59**, 309–337.
32. Painter TS (1921). The Y chromosome in mammals. *Science* **53**, 503–504.
33. Makino S (1956). *A Review of the Chromosome Numbers in Animals.* Tokyo, Hokuryukan.
34. Kottler MJ (1974). From 48 to 46: cytological technique, preconception, and the counting of the human chromosomes. *Bull. Hist. Med.* **48**, 465–502.
35. Harman SO (2004). *The Man who Invented the Chromosome. A Life of Cyril Darlington.* London, Harvard University Press.
36. Belling J (1921). On counting chromosomes in pollen mother cells. *Am. Naturalist.* **55**, 573–574.
37. Levan A (1938). The effect of colchicine on root mitoses in *Allium. Hereditas* **24**, 471–486.
38. Kemp T (1929). Über das verhalten der chromosomen in den somatischen zellen des menschen. *Mikrosk. Anat. Forsch.* **16**, 1–20.
39. Puck TT, Marcus PI and Cieciura SJ (1956). Clonal growth of mammalian cells *in vitro*. Growth characteristics of colonies from single HeLa cells with and without a 'feeder' layer. *J. Exp. Med.* **103**, 273–284.
40. Hughes A (1952). Some effects of abnormal tonicity on dividing cells in chick tissue cultures. *Quart. J. Microscop. Sci.* **93**, 207–220.

41. Makino S and Nishimura I (1952). Water-pretreatment squash technic: a new and simple practical method for the chromosome study of animals. *Stain. Tech.* **27**, 1–7.
42. Andres AH and Jiv BV (1936). Somatic chromosome complex of the human embryo. *Cytologia* **7**, 371–388.
43. Andres AH and Vogel J (1935). Karyological investigation of the embryonal oogenesis in man. *C. R. Acad. Sci. URSS* **4**, 353–354.
44. Zhivago P, Morosov B and Ivanickaya A (1934). Über die einwirkung der hypotonie auf die zellteilung in den gewebkulturen des embryonalen Herzens. *C. R Acad. Sci. URSS* **3**, 385–386.
45. Chrustschoff GK, Andres AH and Ilina-Kakujewa WI (1931). Kulturen von blutleukozyten als methode zum stadium des menslichen karyotypus. *Anat. Anz.* **73**, 159–168.
46. Chrustschoff GK and Berlin EA (1935). Cytological investigations on cultures of normal human blood. *J. Genet.* **31**, 243–261.
47. Andres AH and Navashin MS (1936). Morphological analysis of human chromosomes. *Proc. Maxim Gorky Medico-Genetical Res. Inst.* **4**, 506–524.
48. Andres AH and Shiwago PI (1933). Karyologische studien an myeloischer Leukämie des Menschen. *Folia Haemat.* **49**, 1–20.
49. Medvedev ZA (1969). *The Rise and Fall of TD Lysenko*. New York, Columbia University Press.
50. Soyfer VN (1994). *Lysenko and the Tragedy of Soviet Science*. New Brunswick, Rutgers University Press.
51. Adams MB, Ed. (1990). *The Wellborn Science. Eugenics in Germany, France, Brazil and Russia*. Oxford, Oxford University Press.
52. Penrose LS (1966). Human chromosomes, normal and abnormal. *Proc. Roy. Soc. London B*, **164**, 311–319.
53. Cajal R (1909). *Histologie du Système Nerveux de l'Homme et des Vertébrés*. Paris, A Malone.
54. Moore KL, Ed. (1966). *The Sex Chromatin*. Philadelphia, W.B. Saunders.
55. Barr ML and Bertram EG (1949). A morphological distinction between the neurones of the male and female, and the behaviour of the nucleolar satellite during accelerated nucleoprotein synthesis. *Nature* **163**, 676–677.
56. Wolf U (1974). Theodor Boveri and his book 'On the Origin of Malignant Tumours'. In: German J (Ed.), *Chromosomes and Cancer*, pp. 3–20. New York, John Wiley & Sons Inc.
57. McKusick VA (1960). Walter Sutton and the physical basis for Mendelism. *Bull. Hist. Med.* **35**, 487–497.
58. Crow EW and Crow JF (2003). 100 years ago: Walter Sutton and the chromosome theory of heredity. *Genetics* **160**, 1–4.
59. Koulischer L and Bassleer R (1993). La cytogénétique humaine est née il y a 80 ans à Liège (de Winiwarter, 1912). *Rev. Med. Liège* **48**, 129–136.
60. Leplat G (1960). Éloge académique du Professeur Chevalier Hans de Winiwarter (1875–1949). *Mem. Acad. R. Med. Belg.* **4**, 20–36.
61. Hamerton JL (2001). Painter, Theophilus Schickel, 1889–1969. In: Brenner S and Miller J (Eds) *Encyclopaedia of Genetics*. New York, Academic Press.
62. Matthey R (1949). *Les Chromosomes des Vertébrés*. Lausanne, University of Lausanne.
63. Pathak S (2004). TC Hsu: In memory of a great scientist. *Cytogenet. Genome Res.* **105**, 1–3.

Addendum

Barr ML and Bertram EG (1949). A morphological distinction between neurones of the male and female, and the behaviour of the nucleolar satellite during accelerated nucleoprotein synthesis. *Nature*, **163**, 676–677. Reproduced with permission from Nature Publishing Group.

A Morphological Distinction between Neurones of the Male and Female, and the Behaviour of the Nucleolar Satellite during Accelerated Nucleoprotein Synthesis

Geneticists have long emphasized that "maleness" and "femaleness," so far as chromosome content is concerned, are projected from the fertilized ovum into the morphologically and functionally specialized somatic cells. It appears not to be generally known, however, that the sex of a somatic cell as highly differentiated as a neurone may be detected with no more elaborate equipment than a compound microscope following staining of the tissue by the routine Nissl method.

The observations to be recorded here apply to the cat primarily, since the cat is used routinely in this laboratory for investigations in experimental neurocytology. The nuclei of nerve cells contain a prominent nucleolus which stains readily with such basic dyes as cresyl violet and thionin. The difference in nuclear structure between neurones of adult male and female cats rests on the degree of development of a second body, which is much smaller than the nucleolus. The latter body is more or less intimately associated with the nucleolus and, like the latter, stains well with basic dyes. It has been described by many authors under various names. The term "nucleolar satellite" will be used in this report, in the hope that students of chromosome morphology will not object too strenuously to the use of the word "satellite" in this connexion.

Typically, nerve cells of mature *female* cats contain a well-developed nucleolar satellite which is located, as a rule, immediately adjacent to the nucleolus (Fig. 1–1). A single satellite is usually present; but two may be encountered. The satellite, if more deeply stained than the nucleolus, may be seen in all nerve cells which are sufficiently large to have a prominent nucleolus. More often, the intensity of staining of the nucleolus and its satellite is similar. Under these conditions, the satellite is seen in approximately 30–40 per cent of cells, being invisible when eclipsed by the nucleolus.

As a rule, nerve cells of mature *male* cats (Fig. 1–2) contain a poorly developed nucleolar satellite, seen only infrequently. When visible, it is situated adjacent to the nucleolus and is near the limit of resolution with an oil-immersion objective.

In a small proportion of animals of both sexes, nucleolar satellites of intermediate size are present. These exceptional cases raise several interesting questions which are now under investigation.

The morphological distinction, therefore, between neurones of the mature male and female cat is so clear that sections from the brain, spinal cord or sympathetic ganglia of animals of both sexes may be readily sorted into two groups without prior knowledge of the sex, with only an occasional section remaining in which the distinctive morphological feature is of an intermediate character.

That there should be a morphological difference in the intermitotic nucleus of mature, differentiated cells, according to sex, is not surprising in view of what is known concerning the relation of the nucleolus to the chromosomes. The nucleolar chromosomes are frequently the sex chromosomes.[1] One may postulate that such is the case in the cat. The nucleolar satellite may be derived from the heterochromatin of the sex chromosomes. Further, the cells of the female cat, because of the duplicated X-chromosomes, may be endowed with a greater quantity of nucleolar associated heterochromatin than are the cells of the male cat. Caspersson and Schultz[2] noted a difference in the absorption curves obtained with ultra-violet light, indicating a difference in nucleoprotein content, in cells of male and female *Drosophila*.

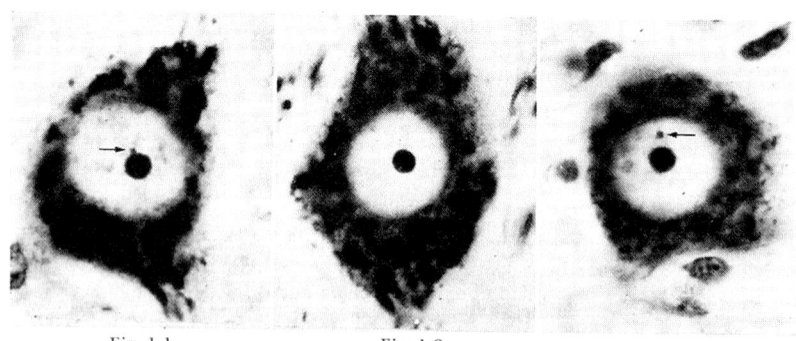

Fig. 1-1   Fig. 1-2   Fig. 1-3

Figure 1-1. Normal motor neuron from the hypoglossal nucleus of a mature female cat showing the usual morphology of the nucleolar satellite (indicated by arrow) in the female. (Cresyl violet stain, ×1400.)

Figure 1-2. Motor neuron from the hypoglossal nucleus of a mature male cat. The nucleolar satellite is absent, the typical condition in the mature male. (Cresyl violet stain, ×1400.)

Figure 1-3. Motor neuron from the hypoglossal nucleus of a mature female cat 108 hours following electrical stimulation of the corresponding hypoglossal nerve for a period of 8 hours. Associated with intense synthesis of cytoplasmic ribose nucleoproteins, the nucleolar satellite (indicated by arrow) tends to move away from the nucleus. (Cresyl violet stain, ×1400.)

These comments suggest the importance of taking the sex into consideration in cytochemical studies of nucleoprotein metabolism.

A preliminary examination of sympathetic ganglia of human males and females indicates that a similar sex difference in nuclear morphology exists in the human. Whether or not sex differences in nuclear structure will be found in a given species would probably be determined, in part, by the relationship of the nucleolus to the sex chromosomes and by the disparity in size and composition between the X- and Y-chromosomes. It is probable that somatic cells of various tissue, characterized by large nucleoli, will display similar distinctive nuclear differences according to the sex.

We wish to emphasize that these observations apply to mature animals. The influence of the age factor and other aspects of the nucleolar satellite under normal and experimental conditions will be published in later reports.

It is of interest to experimental neurocytologists that the position of the satellite relative to the nucleolus is a useful aid in assessing the physiological state of the cell. In a series of experiments, to be reported in detail elsewhere, depletion of the Nissl material of motor neurones was produced by prolonged electrical stimulation of the hypoglossal nerve. The nucleolus enlarges during the recovery phase, coincident with the re-appearance of abundant Nissl material in the cytoplasm. This observation is in agreement with the views of Caspersson and his co-workers, who regard the nucleolus and the nucleolar associated chromatin as instrumental in the synthesis of ribose nucleoprotein, an important component of the Nissl substance (see Hydén[3] for references). It is of particular interest in the present connexion that the satellite moves away from the nucleolus during the period of intense ribose nucleoprotein synthesis (Fig. 1-3). The satellite may, in these circumstances, lie in contact with the nuclear membrane. The movement of the nucleolar satellite may be passive, resulting from the outpouring of materials (nucleotides or nucleic acids?) from the region of the nucleolus. On the other hand, complex factors, such as forces of an electrical nature, may be at work. In any case, the position of the satellite is another item for observation, in addition to the appearance of the Nissl substance, in attempting to assess the physiological state of the neurone under experimental conditions. This criterion may be applied to best advantage, of course, only to the cells of female cats. The nucleolar satellite is occasionally found free in the nucleoplasm in control cells. This is regarded as a further indication of the variation in the physiological state of members of a nerve-cell population under normal conditions.

It is hoped that this brief preliminary report may encourage closer attention to the nucleolar satellite in the abundant material available to laboratories of neuropathology in which the Nissl method is constantly in use as a routine staining technique. The possibility that fundamental alterations in the nucleolar associated chromatin may have an important bearing on malignancy gives added interest to these observations.

These observations were made in the course of experiments on the effect of activity on the neurone being done in this laboratory for the Institute of Aviation Medicine, Royal Canadian Air Force. The senior author wishes to thank various members of the staffs of the Department of Neuropathology, University of Toronto, and of the Montreal Neurological Institute, for very helpful discussions.

MURRAY L. BARR
EWART G. BERTRAM

Department of Anatomy,
University of Western Ontario,
London, Canada.
March 3.

1. Gates, R. R., *Bot. Rev.*, 8, 337 (1942).
2. Caspersson, T., and Schultz, J., *Proc. U.S. Nat. Acad. Sci.*, 26, 507 (1940).
3. Hydén, H., *Symp. Soc. Exp. Biol.*, 1, 152 (1947).

CHAPTER 2

46, not 48: the discovery of the human chromosome number, 1956

ON 22ND DECEMBER, 1955, AT 2 a.m., Dr Joe Hin Tjio, working at his microscope in the laboratory of Albert Levan in the Institute of Genetics of the University of Lund in Sweden, noted his finding that the diploid human chromosome number was 46, not 48 as had been generally supposed for the previous 30 years. We know the date and time of this discovery with unusual precision since Tjio, a passionate photographer, later sent copies of his first photomicrographs to friends and colleagues around the world, inscribed in a variety of languages, as can be seen in Figure 2-1.

Confirmation of these precise circum-

Fig. 2-1 Human karyotype showing 46 chromosomes (as sent by Tjio to colleagues). The inscription in the lower left-hand corner states: 'Human cell with 46 chromosomes observed 1955 on December 22nd at 2.00 am.' (Courtesy of Professor Patricia Jacobs.)

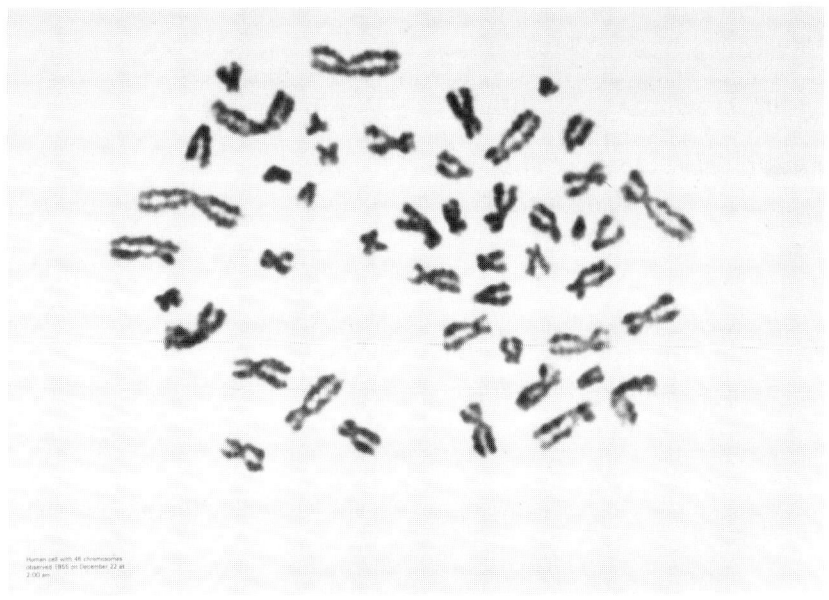

JOE HIN TJIO (1919–2001)

Joe Hin Tjio was born in Indonesia, being of Chinese descent. Trained as an agronomist, he suffered greatly both during and after World War II, being imprisoned and tortured first by the Japanese and later by the Indonesian government for left-wing activities.

Leaving Indonesia, he moved first to the Netherlands, then to the Plant Breeding Institute at Svalöf, near Lund, Sweden, where he began work on plant chromosomes and first collaborated with Albert Levan. He maintained his Swedish links after taking an appointment in Zaragoza, Spain, spending considerable periods in Lund up to the time of the discovery of the human chromosome number.

At the end of 1956 he was persuaded by Theodore Puck to come to the University of Denver to develop human cytogenetics, though with some reluctance in relation to his early life, since the United Sates was then in its 'McCarthyist' phase of anticommunist persecution. Before the 1960 Denver conference he had again moved to the National Institutes of Health, Washington, where he remained for the rest of his career.

From an early age, and throughout his life, Tjio was a passionate photographer, an interest probably acquired from his father, a professional photographer. Many memories and anecdotes are recalled by colleagues internationally, but while there are brief published notes on his life[19,20] there is at present no full account. (Photograph reprinted from *The Cells of the Body* by Henry Harris, © Cold Spring Harbor Laboratory Press, with permission.)

---

stances comes from the account of Professor Maj Hultén,[1] who was working for a short period in Lund at the end of 1955, while a genetics student in Stockholm. She gives a vivid picture of the Lund laboratory at that time, where many workers began their research at the end of the day when teaching commitments had been finished.

After dark the Institute changed scene completely. Now, when the students had left, the scientific work could begin. This was probably the happiest working time in my life.

Tjio's work seems to have been particularly intense and nocturnal. Based primarily at this time at the University of Zaragoza in Spain, he had been a visiting scientist in Lund with Albert Levan for several periods over the previous years and had probably become involved in this project in part

ALBERT LEVAN (1905–1998)

Albert Levan was born in Gothenburg, Sweden, and trained in botany at the University of Lund, graduating in 1927. Working at the Svalöf Institute of Plant Breeding near Lund from the 1930s, he made extensive studies of plant chromosomes, developing techniques assessing chromosome damage by toxic agents using the dividing group tip cells of *Allium*, including analysis of the effects of colchicine on mitosis.

Impressed by the similarity of these changes to alterations in cancer cells, he changed his research field to cancer cytogenetics and subsequently moved back to the University of Lund, where he developed a cancer chromosome laboratory and collaborated widely with other workers, notably in the United States, where he spent extended periods with T Hauschka (Philadelphia) and J Biesele (New York).

Having, with Joe Hin Tjio, established the normal human chromosome number largely as a baseline for his cancer studies, he pursued these for the rest of his career, leaving a continuing and flourishing tradition of cancer cytogenetics in Lund. While the film made by Levan's Lund colleagues provides a valuable personal record, only brief accounts on his life are available in English at present.[21] (Photograph reprinted from *The Cells of the Body* by Henry Harris, © Cold Spring Harbor Laboratory Press, with permission.)

because of his technical and photographic skills. Although in Lund for much of the summer of 1955, he returned from Spain only in December, shortly before the discovery, and possibly in response to a message that the monolayer embryonic cell cultures prepared by Professor Rune Grubb, head of the university microbiology department, were ready to analyse.

Maj Hultén was one of the first people to see Tjio's preparations and she describes it graphically in her article.

I was walking in the culvert linking the Institute to the Animal House, carrying my mouse cases. It was late at night the day before Christmas Even, on December 23, 1955, when I suddenly heard the clapping (and echoing) sound of clogs behind me, and a heavy hand landed on my left shoulder. I got mighty afraid, but recognizing it to be the diminutive Chinese visiting scientist, Joe Hin Tjio, I wondered what on earth this was all about. 'I can see that you are equally kind to everybody around here. Would you like to come to my room? I have got something interesting to show you', he stuttered. 'Yes, please', I found myself answering.

Peering down the microscope situated on the bench to the right in Tjio's office cum lab, I was amazed to see the human chromosomes well spread out and separated from each other; and when Tjio demanded: 'Count', I did so. My first comment was 'you've lost two', but then in metaphase after metaphase there could be no doubt, the chromosome number was 46.

The full significance of the discovery is best appreciated by reading the paper by Tjio and Levan,[2] reproduced here, that appeared in *Hereditas* the following April, 1956, having been submitted on 26th January. The quality and clarity of the photomicrographs, the large number of cells analysed and the fact that virtually all (all except four) cells showed 46 chromosomes, are in contrast to the uncertainties and ambiguity of most of the previous studies, and left the international cytogenetics research community in no doubt that a major reassessment of previous conclusions was needed. Inevitably, a few remained unconvinced;

CHARLES E FORD (1912–1999)

Born in Cheshire, England, but moving south with his family to Slough, near London, Charles Ford graduated in botany from King's College, London in 1931, and did his PhD on the chromosomes of *Oenothera*, the evening primrose. He then moved to Ceylon to lead research into rubber genetics, interrupted by wartime military service, and in 1946 spent time at Chalk River, Canada, studying radiation damage to chromosomes, using the *Vicia faba* root tip model.

Returning to Britain in 1949 he joined the new Medical Research Council (MRC) Radiobiology Unit at Harwell, where his key work was done. Turning from plant to mouse chromosomes as a model for radiation damage he developed, with John Hamerton, the T6 translocation which provided a marker that could distinguish whether bone marrow regeneration resulted from host or transplanted marrow, a technique that later proved vital in human marrow transplantation after radiotherapy.

Ford's involvement in human cytogenetics began with the confirmation of the chromosome number in 1956, and continued with the development of bone marrow cytogenetic techniques and the identification of sex chromosome abnormalities (see Chapters 3 and 4), but he never became fully involved in medical aspects, even after his group moved to Oxford in 1971.

Charles Ford's ebullient personality, enthusiastic lecturing style and numerous foibles, together with his meticulous microscopy skills, greatly endeared him to successive students, though his perfectionist approach to writing meant that too much work stayed unpublished. Numerous anecdotes, scientific and non-scientific, exist, some of which can be found in published memoirs,[22–24] as well as in interviews with his colleagues. (Photograph reproduced from *Cytogenetics and Cell Genetics*, Vol. 20, 1978, courtesy of S. Karger AG, Basel.)

JOHN L HAMERTON (born 1929)

John Hamerton was born in London (UK) and graduated in zoology at Imperial College, London University. In 1951 he began working with Charles Ford at Harwell on mouse chromosomes in relation to radiation damage, also spending time initially with the Slizynskis in Edinburgh, from where a number of scientists in the MRC group would soon relocate to Harwell. His involvement (and that of Ford) in human cytogenetics only began in 1956, with their work on determination of the meiotic human chromosome number.

After leaving Harwell in 1956 and spending three years as a scientist at the British Museum, Hamerton joined Paul Polani at Guy's Hospital, London, in 1960, soon after the discovery of the first sex chromosome abnormalities; here he set up a comprehensive clinical cytogenetics laboratory. His 1971 book, *Human Cytogenetics*,[25] is a definitive synthesis of the field prior to chromosome banding.

In 1969 John Hamerton became head of human genetics at the University of Winnipeg, Canada, developing a series of large-scale population chromosome surveys, pre-natal diagnosis trials and hybrid cell studies in gene mapping, and playing a major role in the development and coordination of genetic services across Canada. (Photograph courtesy of Professor John Hamerton.)

TT Puck, in his autobiographical essay (see Chapter 7) records one (unnamed) American authority as commenting:

Isn't it wonderful that science in our time is so free that even nonsense like this can get a hearing?

The First International Human Genetics Congress, held later in 1956 in Copenhagen,[3] provided a further opportunity for Tjio to give a demonstration of the findings (Figure 2-2) and to remove any remaining doubts.

Independent published confirmation rapidly followed, Ford and Hamerton from Harwell, England, reporting their findings on meiotic chromosomes from human testicular tissue in the same year.[4] It has been suggested that this was an independent discovery, but this is not the case. Not only do Ford and Hamerton cite the Tjio and Levan paper, but interviews in 2004 with Charles Ford's colleagues, John Hamerton and Ted Evans, confirmed that there were close contacts between the groups at Harwell and Lund over this time. As John Hamerton recalls:

In the early fifties it was clear that man had 48 chromosomes. It was, I think, in 1955 that we had a visit from a urologist from Oxford, who was just visiting as people do, and he said to Charles 'I can get you some human material if you want'. Charles said 'sometime when we've got time we'll

Fig. 2-2 Joe Hin Tjio displaying preparations of the human chromosomes at the First International Human Genetics Congress, Copenhagen. (Courtesy of Professor David Harnden.)

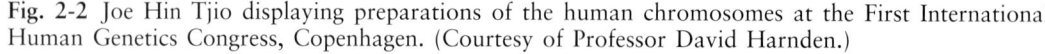

have human material'. We had the techniques, we had the ability to look at meiotic chromosomes which not many other labs did.

Hsu had looked and he got 48, everybody seemed to get 48, and then we got this offer of human material and we said, when we've got time we'll do that, we are very busy, there didn't seem there was a rush. And then on the grapevine we heard something about Tjio and Levan's work before it was published and so Charles got back to the urologist and we very quickly got testicular material at the end of '55 and did the studies and confirmed Tjio and Levan's counts.

The possibility was raised, though, that some form of polymorphism might exist for the human chromosome number, as it does in some other mammals, or even that the number might differ between races.⁵ The development of skin fibroblast cultures allowing samples to be sent long distances allowed this to be examined by David Harnden,⁶ who had succeeded John Hamerton as colleague of Charles Ford. Harnden has described this study in his historical review of human cytogenetics⁷ and in a 2004 interview:

Perhaps the most unusual was an attempt to determine whether reports of variation in the normal chromosome number could be substantiated. There had been some discussion about such variations in chromosome numbers in different human populations, but nothing like that was turning up, so I decided that it might just be worth looking at the chromosomes of a native Australian. With the help of a colleague in Adelaide and the Flying Doctor Service, I was able to get a skin biopsy from an assuredly 'full-blooded' aborigine. The sample reached me at Harwell from the 'Outback' in 4 days. It grew beautifully but the chromosomes were quite normal. A disappointment, I suppose, but still fun to do.

The discovery of the human chromosome number was clouded at the time, and for many years afterwards, by a dispute between Tjio and Levan as to their respective contributions, and as to who should be first author on the paper. This is particularly sad since there is no doubt that the contribution of both was essential. Since both Tjio and Levan are now dead it seems important to record the true situation, as far as is still possible, especially as Levan's role has been downplayed in some American accounts.⁸ Maj Hultén's paper¹ records the memories of one who was there at the time, albeit briefly, while I was able to discuss the work with colleagues in Lund in October, 2004.⁹ There is general agreement on most of the main points: the specific initial observations in December 1955 were made by Tjio, but the project was initiated and directed by Levan, who was present throughout the critical period and whose funding supported Tjio as visiting worker, the aim of the work being primarily to provide a normal control basis for Levan's major research programme on cytogenetic abnormalities in human cancers. It is likely that Tjio's traumatic early life (see box) and resulting over-sensitivity made what should have been a temporary disagreement into a permanent breach, though Levan always referred warmly and respectfully to Tjio's role, as can be seen in a filmed interview made late in his life in Lund. (This interview, in Swedish, with Professor Bengt-Olle Bengtsson, is held at the Lund University Genetics Department.) Both Tjio and Levan wrote brief (and separate) retrospective notes on the discovery in 1978.¹⁰,¹¹

Fifty years later, with the recognition that this discovery was the starting point for modern human cytogenetics, we are more concerned with two important scientific questions: first, why did the discovery occur in Lund, and second, how did it happen that a series of able investigators over the previous 30 years had agreed in reaching the fundamentally wrong conclusion that the human chromosome number was 48?

The first of these questions is more easily answered than the second. Lund was (and is) a major scientific centre and Albert Levan's cancer chromosome laboratory in the Institute of Genetics (Figures 2-3 and 2-4), headed by Arne Muntzing, was well equipped for cytogenetics and familiar with the technological advances described in the previous chapter. Both Levan and Tjio had begun their careers by extensive work on plant cytogenetics and retained close links with the neighbouring Plant Breeding Institute at Svalöf, where Levan had previously been one of the first to use colchicine in cytogenetic analysis.[12] Tjio's technical ability and photographic skills have been mentioned, but a further factor was the availability of human cultured embryonic cells, Sweden being one of the few countries at the time where abortion was legal, though only on a restricted basis. A key element in the project was the provision of cultured fetal lung fibroblasts

Fig. 2-3 Albert Levan with his fellow professors Arne Münzing (left) and Åke Gustavsson (right) at the Lund Institute of Genetics. (Courtesy of Reinards Hochberg, University of Lund.)

Fig. 2-4 (a) The Institute of Genetics, Lund. (b) Plaque at the Institute entrance commemorating the discovery of the human chromosome number. (Courtesy of Dr Ulf Kristoffersson, Lund.)

(a)

(b)

by Rune Grubb, Professor of Microbiology in Lund. Indeed, the timing of their availability would have been a determining factor in Tjio's return to Lund from Spain in December 1955, allowing him to undertake the microscopic work within a short period of time. It has been queried why Grubb was not a co-author on the paper, but his help was acknowledged and he probably considered this appropriate.

There is, however, an additional reason why the discovery of the human chromosome number occurred in Lund and this, while mentioned in the 1956 paper, has not been adequately recognised. This is that work in the Lund Institute of Genetics during the previous two years had already shown that the 46 chromosome number was likely, and that Levan probably knew this at the time of the subsequent findings.

During 1954, Dr Eva Melander, working with Levan, and her husband Dr Yngve Melander, made chromosome preparations from cultured fetal liver provided by their gynaecology colleague Dr Stig Kullander, and observed 46 chromosomes. I was able to discuss this work with them in Lund in October 2004 and Dr Eva Melander kindly gave me one of her photomicrographs, never published and apparently not seen by others during the subsequent 50 years (Figure 2-5). The results were shared with Levan at the beginning of 1955 and Eva Melander considers that Levan was convinced of the correctness of the 46 number before he left to work at the Sloan-Kettering Institute, New York, during Spring 1955. This suggests that the subsequent work of Tjio and Levan may have been done in the expectation of finding 46 chromosomes, rather than being mentally hindered by the established view of 48 being the correct number.

On the other hand, Levan's 1956 paper on his Sloan-Kettering tumour work,

Fig. 2-5 Human chromosome preparation by Drs Eva and Yngve Melander, Lund, showing 46 chromosomes, dated May 1954. (Courtesy of Dr Eva Melander.)

published in *Cancer*[13] and received by the journal on 1st December, 1955, notes that a normal cultured fibroblast line *'gave the expected normal human chromosome number'*. That this indicates 48 chromosomes is clear from a reference to Hsu's 1952 paper[14] in the following sentence, and from Levan's statement that *'the present fibroblast analysis is roughly in agreement with Hsu's idiogram'*. One would not have expected this clear statement from a cautious worker such as Levan, if he was already convinced that the true human chromosome number was 46.

It is not entirely clear why the work of the Melanders was not published, but in our 2004 discussion Eva Melander emphasised that she felt the quality was not adequate; once the undoubtedly superior preparations of Tjio had been made, it would have been less satisfactory to publish the earlier preparations. However this may be, the work of Eva and Yngve Melander deserves recognition as a key step in human cytogenetics.

As to why so many other workers had consistently, even persistently, concluded that the human diploid chromosome number was 48, this provides some important general lessons on the fallibility of science and scientists that remain as relevant today as 50 years ago. The topic has been well studied, albeit 30 years ago, from the perspective of a science historian, Kottler,[15] and again recently by Martin[16] in the general context of how and why objects are counted; it has also been debated by scientists such as TC Hsu, who understandably still felt upset many years later that he had failed to reach the correct conclusion despite having done more than almost anyone to make it possible. Hsu, in his 1979 book[17] writes:

When Malcolm Kottler was doing his research on the history of human karyology, he asked me a number of questions. I mentioned to him that not finding the correct human chromosome number was a sore point in my career. Kindly he consoled me by writing:

'*Not that my opinion matters, but I don't think you should consider your "confirmation" of the old result, 2N = 48, a sore point in your career. Though both tissue culture and hypotonic pre-treatment were old, even if rarely used, methods in animal cytology, it seems clear that your reintroduction of both techniques in the early 1950s was a turning point of critical significance.*'

In my letter to answer some more of his questions, I commented on that point:

'*I am not belittling my contribution. I do believe the method I introduced opened the door to the booming progress of the present day human cytogenetics and mammalian genetics. But after returning an interception for 40 yards and fumbling the ball at the three-yard line, one always laments that he could have made a touchdown. I sincerely thank you for giving credit to that futile 40 yard run, for it was still the play that saved the game!*'

Hsu attributes his error both to technological factors and to his respect for Painter, who had been his mentor and who was president of his university (University of Texas): '*It was unthinkable that Painter could be wrong.*' Looking at the data in Hsu's paper makes it clear that the problem was more one of interpretation than of technology, the cell counts (Table 2-1) showing a sharply skewed distribution with the great majority scored as 48, rather than the wider spread of counts to be expected if the quality of the preparations

TABLE 2-1
CELL COUNTS OF HUMAN CHROMOSOMES

Number of cells	Chromosome number
1	44
5	45
11	46
11	47
91	48
4	49

(From TC Hsu, 1952[14])

had been the determining factor. As Hsu himself notes:

We 'knew' that the diploid number of man was 48, but I thought that the superior morphology in my slides could help me describe them in more detail. This preconception of 48 as the diploid number made the job very difficult, because in many cells I had difficulty in getting the count to equal 48. In order to 'force' a 48 count, I decided that some biarmed chromosomes with stretched centromeric regions must be two separate acrocentric chromosomes.

Hsu later had the opportunity to examine some of Painter's original preparations and was amazed that it had been possible to draw any conclusions at all from them, so crowded and tangled were the chromosomes. The same could be said of almost all other studies prior to those of Hsu himself.

Thus, the reasons underlying the failure to discover the correct human chromosome number until the work in Lund were, as noted by Kottler:

1. Inadequate technology to allow accurate counts, until the successive developments outlined in Chapter 1 had occurred.
2. Interpretation of the data well beyond what it could be reasonably expected to have allowed.
3. The subconscious influence of the earlier studies on later investigators when interpreting their own data, so that their conclusions were based largely on what they expected to find.

In addition, it must be remembered that, as noted in Chapter 1, many early studies were of meiotic chromosomes, with the diploid number inferred from this. Early separation of the XY bivalent might often have given the appearance of 24 chromosomes, making it natural to infer that the diploid number would be 48.

Tjio and Levan reached the correct conclusion partly because their own technical skills and the use in combination of previous technological advances made their results completely unambiguous, but also because the previous work of the Melanders had removed, or at least shaken the expectation that the human chromosome number must be 48.

Current research workers will do well to recognise that these problems are pervasive in science and that some scientific advances are simply not possible until the technology involved reaches a particular stage. Many new technologies in their early stages, including those of molecular genetics, are just as dependent on interpretation, visual or perceptive, as were those facing the early cytogeneticists, but it is always difficult to admit in print, as did Winiwarter, that one cannot make a conclusive and exact result.

In such a necessarily imprecise situation there is a particular risk of trying to fit conclusions in with previously accepted results, which may on occasion prove to be erroneous. The consequence of this is heightened today by publication bias, where studies failing to confirm or disagreeing with an initial observation are likely to be relegated to lower impact journals or not published at all.

The 1956 publication of the human chromosome number has rightly been defined as the foundation not just of human cytogenetics, but also as a key factor in the development of medical genetics. It gave workers in this field, as stated by Victor McKusick in his historical review,[18] their own organ in the form of the chromosome, as a counterpart to the body systems of other medical specialties, in a way that earlier theoretical studies, or the work on inborn errors of metabolism by Garrod and his successors, had not done. However, neither Tjio nor Levan, nor indeed Lund as a centre, were to play a major role in this development of clinical cytogenetics. Tjio moved to America (initially to Denver, with Theodore Puck, then to the National Institutes of Health), where his later contributions never matched the significance of his 1956 discovery. Levan, now having achieved the aim of establishing a normal cytogenetic baseline for his cancer cytogenetics research, returned to this and developed a distinguished record in the field that his successors in Lund continue to this day. Levan continued his cancer cytogenetics work until the end of his long life; in the filmed interview he states: 'For 50 years I have looked at chromosomes every day; I regard them as my friends.'

References

1. Hultén M (2002). Numbers, bands, and recombination of human chromosomes: Historical anecdotes from a Swedish student. *Cytogenet. Genome Res.* **96**, 14–19.
2. Tjio JH and Levan A (1956). The chromosome number of man. *Hereditas* **42**, 1–6.
3. Kemp T (1956). Über das verhalten der chromosomen in den somatischen zellen des menschen. *Mikrosk. Anat. Forsch.* **16**, 1–20.
4. Ford CE and Hamerton JL (1956). The chromosomes of man. *Nature* **178**, 1020–1023.
5. Kodani M (1958). The supernumerary chromosome of man. *Am. J. Hum. Genet.* **10**, 125–140.
6. Harnden DG (1960). A human skin culture technique used for cytological examination. *Br. J. Exper. Pathol.* **41**, 31–37.
7. Harnden DG (1996). Early studies on human chromosomes. *BioEssays* **18**, 163–168.
8. Barnes B (2001). Joe Hin Tjio; pioneer in human genetics. *Washington Post*, 4th December, 2001.
9. Harper PS (2006). The discovery of the human chromosome number in Lund, 1955–6. *Hum. Genet.* in press.
10. Tjio JH (1978). The chromosome number of man. *Am. J. Obstet. Gynecol.* **130**, 723–724.
11. Levan A (1978). The background to the determination of the human chromosome number. *Am. J. Obstet. Gynecol.* **130**, 725–726.

12. Levan A (1938). The effect of colchicine on root mitoses in *Allium. Hereditas* **24**, 471–486.
13. Levan A (1956). Chromosome studies of some human tumors and tissues of normal origin, grown in vivo and in vitro at the Sloan-Kettering Institute. *Cancer* **9**, 648–663.
14. Hsu TC (1952). Mammalian chromosomes in vitro. 1. The karyotype of man. *J. Hered.* **43**, 172.
15. Kottler MJ (1974). From 48 to 46: cytological technique, preconception, and the counting of the human chromosomes. *Bull. Hist. Med.* **48**, 465–502.
16. Martin A (2004). Can't any body count? Counting as an epistemic theme in the history of human chromosomes. *Social Stud. Sci.* **34**, 1–26.
17. Hsu TC (1979). *Human and Mammalian Cytogenetics. An Historical Perspective.* New York, Springer.
18. McKusick VA (2001). A history of medical genetics. In: *Emery and Rimoin's Principles and Practice of Medical Genetics.* London, Churchill-Livingstone, pp. 3–6.
19. McManns R (1997). Photographer, pioneer, polyglot NIDDK's Tjio ends distinguished scientific career. NIH Record. www.nih.gov/news/record/02–11–97/story01.htm.
20. Hultén M (2003). Tjio, Joe Hin. *Encyclopedia of the Human Genome.* Oxford, Macmillan, pp. 1969–1970.
21. Hultén M and Fredga K (2003). Levan, Albert. *Encyclopedia of the Human Genome.* Oxford, Macmillan, pp. 1094–1095.
22. Kent-First M (1999). Charles Edmund Ford Ph.D., FRS. In loving memory of my mentor and friend. *Cytogenet. Cell Genet.* **85**, 193–195.
23. Lyon MF (2001). Charles Edmund Ford. *Biog. Mems. Fell. R. Soc. Lond.* **47**, 189–201.
24. Gropp A, Pearson PL and Klinger HP (1978). Charles E Ford. *Cytogenet. Cell Genet.* **20**, v–x.
25. Hamerton JL (1971). *Human Cytogenetics*, Vols 1 and 2. New York, Academic Press.

Addendum 1

Tjio JH and Levan A (1956). The chromosome number of man. *Hereditas*, **42**, 1–6. Reproduced with permission from the Mendelian Society of Lund.

THE CHROMOSOME NUMBER OF MAN

By JOE HIN TJIO and ALBERT LEVAN

ESTACION EXPERIMENTAL DE AULA DEI, ZARAGOZA, SPAIN, AND CANCER CHROMOSOME
LABORATORY, INSTITUTE OF GENETICS, LUND, SWEDEN

WHILE staying last summer at the Sloan-Kettering Institute, New York, one of us tried out some modifications of Hsu's technique (1952) on various human tissue cultures carried in serial *in vitro* cultivation at that institute. The results were promising inasmuch as some fairly satisfactory chromosome analyses were obtained in cultures both of tissues of normal origin and of tumours (LEVAN, 1956).

Later on both authors, working in cooperation at Lund, have tried still further to improve the technique. We had access to tissue cultures of human embryonic lung fibroblasts, grown in bovine amniotic fluid; these were very kindly supplied to us by Dr. RUNE GRUBB of the Virus Laboratory, Institute of Bacteriology, Lund. All cultures were primary explants taken from human embryos obtained after legal abortions. The embryos were 10—25 cm in length. The chromosomes were studied a few days after the *in vitro* explantation had been made.

In our opinion the hypotonic pre-treatment introduced by HSU, although a very significant improvement especially for spreading the chromosomes, has a tendency to make the chromosome outlines somewhat blurred and vague. We consequently tried to abbreviate the hypotonic treatment to a minimum, hoping to induce the scattering of the chromosomes without unfavourable effects on the chromosome surface. Pre-treatment with hypotonic solution for only one or two minutes gave good results. In addition, we gave a colchicine dose to the culture medium 12—20 hours before fixation, making the medium 50×10^{-7} mol/l for the drug. The colchicine effected a considerable accumulation of mitoses and a varying degree of chromosome contraction. Fixation followed in 60 % acetic acid, twice exchanged in order to wash out the salts left from the culture medium and from the hypotonic solution that would otherwise have caused precipitation with the orcein. Ordinary squash preparations were made in 1 % acetic orcein. For chromosome counts the squashing was made very mild in order to keep the chromosomes in the metaphase groups. For idiogram studies a more thorough squashing was preferable. In many cases single cells were squashed under the microscope by a slight pressure of a needle. In such cases it was directly observed that no chromosomes escaped.

THE CHROMOSOME NUMBER

With the technique used exact counts could be made in a great number of cells. Figs. 1 *a* and *b* represent typical samples of the appearance of the chromosomes at early metaphase (*a*) and full metaphase (*b*), showing the ease with which the counting could be made. In Table 1 the numbers of counts made from the four embryos studied are recorded.

TABLE 1. *Number of exact chromosome counts made.*

Embryo No.	Number of cultures	Number of counts
1	5	15
2	10	98
3	3	119
4	4	29
Total	22	261

We were surprised to find that the chromosome number 46 predominated in the tissue cultures from all four embryos, only single cases deviating from this number. Lower numbers were frequent, of course, but always in cells that seemed damaged. These were consequently disregarded just as the solitary chromosomes and the groups with but a few chromosomes, which were frequent. In some doubtful cases the numbers 47 and 48 were counted (in four cases not included in the table). This may be due to one or two solitary chromosomes having been pressed into a 46-chromosome plate at the squashing. It is also possible that deviating numbers may originate through non-disjunction, thus representing a real chromosome number variation in the living tissue. This kind of variation will probably increase as a consequence of the change in environment for the tissue involved in the *in vitro* explantation. Hsu (1952) reports a certain degree of such variation in his primary cultures. LEVAN (1956), studying long-carried serial subcultures, found hypotriploid stemline numbers in two of them, and a near-diploid number in a third culture. In this culture one cell with 48 chromosomes was analysed. Naturally, at that time, this was thought to represent the normal diploid number.

CHROMOSOME MORPHOLOGY

Some data on the chromosome morphology of the 46 human chromosomes will be communicated here. The detailed idiogram analysis will be postponed, however, until we are able to study individuals of known sex, the sex of the present embryos being unknown. The comparative study of germline chromosomes in spermatogonial mitoses constitutes an urgent supplement to the present work.

In Fig. 2 four cells are analysed ranging from late prophase (a) to late c-metaphase (d). The chromosomes of metaphases with moderate colchicine contraction vary in length between 1 and 8 μ (Fig. 2 b), but the entire range of variation of Fig. 2 is from 1 to 11 μ. The chromosome morphology is roughly concordant with the observations of earlier workers, as, for instance, the idiogram of Hsu (1952). The chromosomes may be divided into three groups: M chromosomes (median-submedian centromere; index long arm : short arm 1—1,9), S chromosomes (subterminal centromere; arm index 2—4,9), and T chromosomes (nearly terminal centromere; arm index 5 or more).

The M and S chromosomes are present in about equal numbers (twenty of each), while six T chromosomes are found. The classification of the three groups is arbitrary, of course, since gradual transitions of arm indices occur between the three groups. Certain submedian M chromosomes are hard to distinguish from some of the S chromosomes, and the most asymmetric S chromosomes approach the T group.

The chromosomes are easily arranged in pairs, but only certain of these pairs are individually distinguishable. Thus, the M chromosomes include the three longest pairs, which can always be identified. The two longest pairs are different: the second having a decidedly more asymmetric location of its centromere. The two or three smallest M pairs are also recognizable. Between the three longest and the three shortest pairs there are four intermediate pairs that cannot be individually recognized.

The S chromosomes are hardly identifiable, since they form a series of gradually decreasing length. The largest pair, however, is characteristic. Certain chromosomes were seen to have a small satellite on their short arms. Secondary constrictions, too, have been observed now and then, so that it may be hoped that the detailed morphologic study will lead to the identification of more chromosome pairs. The T chromosomes are recognizable: they constitute three pairs of middle-sized chromosomes. Unlike the mouse chromosomes, the human T chromosomes evidently have a small shorter arm.

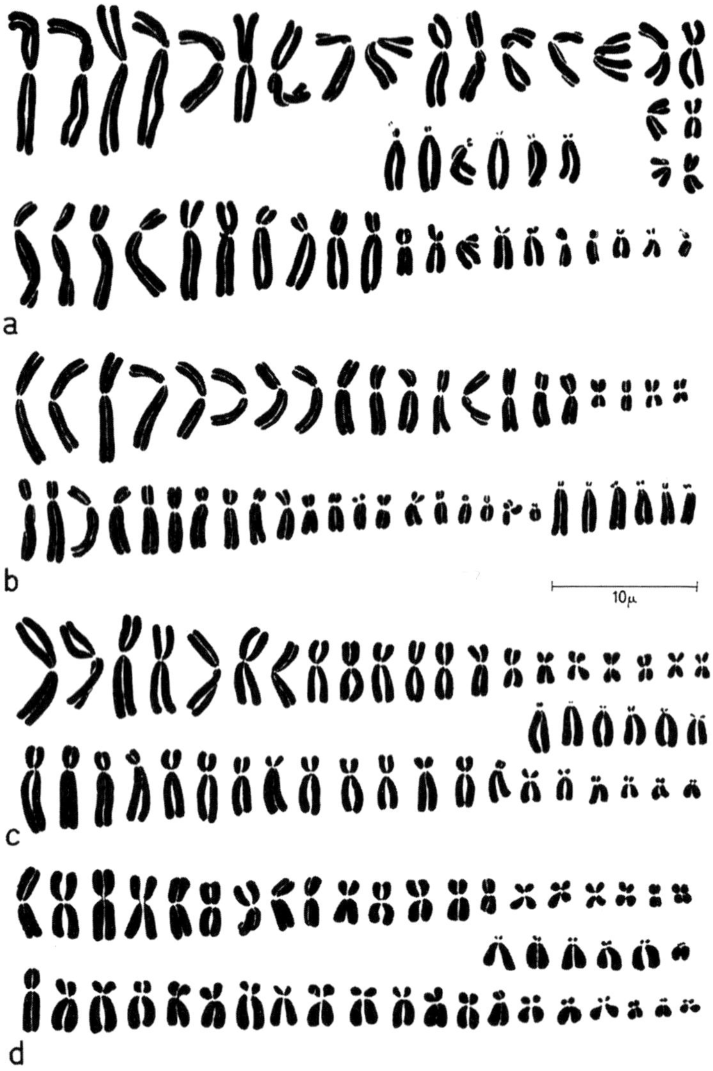

Fig. 2. Four idiogram analyses of human embryonic lung fibroblasts grown *in vitro*. The chromosomes have been grouped in three classes: M (top row), S (bottom row), and T (in between, except in *b*, where T is at the end of the S row). Within each class the chromosomes have been roughly arranged in diminishing order of size. — ×2400.

CONCLUSION

The almost exclusive occurrence of the chromosome number 46 in one somatic tissue derived from four individual human embryos is a very unexpected finding. To assume a regular mechanism for the exclusion of two chromosomes from the idiogram at the formation of a certain tissue is unlikely, even if this assumption cannot be entirely dismissed at this stage of inquiry. Our experience from one somatic tissue in mice and rats, *viz.*, regenerating liver, speaks against this assumption. The exact diploid chromosome set was always found in regenerating liver.

After the conclusion had been drawn that the tissue studied by us had 46 as chromosome number, Dr. EVA HANSEN-MELANDER kindly informed us that during last spring she had studied, in cooperation with Drs. YNGVE MELANDER and STIG KULLANDER, the chromosomes of liver mitoses in aborted human embryos. This study, however, was temporarily discontinued because the workers were unable to find all the 48 human chromosomes in their material; as a matter of fact, the number 46 was repeatedly counted in their slides. We have seen photomicrographs of liver prophases from this study, clearly showing 46 chromosomes. These findings suggest that 46 may be the correct chromosome number for human liver tissue, too.

With previously used technique it has been extremely difficult to make counts in human material. Even with the great progress involved in HSU's method exact counts seem difficult, judging from the photomicrographs published (HSU, 1952 and elsewhere). For instance, we think that the excellent photomicrograph of HSU published in DARLINGTON's book (1953, facing p. 288) is more in agreement with the chromosome number 46 than 48, and the same is true of many of the photomicrographs of human chromosomes previously published.

Before a renewed, careful control has been made of the chromosome number in spermatogonial mitoses of man we do not wish to generalize our present findings into a statement that the chromosome number of man is $2n=46$, but it is hard to avoid the conclusion that this would be the most natural explanation of our observations.

Acknowledgements. — We wish to express our sincere thanks to the Swedish Cancer Society for financial support of this investigation, and to Dr. RUNE GRUBB for supplying us with tissue cultures.

SUMMARY

The chromosomes were studied in primary tissue cultures of human lung fibroblasts explanted from four individual embryos. In all of them the chromosome number 46 was encountered, instead of the expected number 48. Since among 265 mitoses counted all except 4 showed the number 46, this number is characteristic of the tissue studied. The possible bearing of this result on the chromosome number of man is discussed.

Institute of Genetics, Lund, January 26, 1956.

Literature cited

DARLINGTON, C. D. 1953. The facts of life. — London. 467 pp.
HSU, T. C. 1952. Mammalian chromosomes *in vitro*. — The karyotype of man. — J. Hered. 43: 167—172.
LEVAN, A. 1956. Chromosome studies on some human tumors and tissues of normal origin, grown *in vivo* and *in vitro* at the Sloan-Kettering Institute. — Cancer (in the press).

Fig. 1. Colchicine-metaphases of human embryonic lung fibroblasts grown *in vitro*. (a) Early metaphase, (b) full metaphase. The two cells are from embryos 2 and 3 (Table 1) respectively – ×2300.

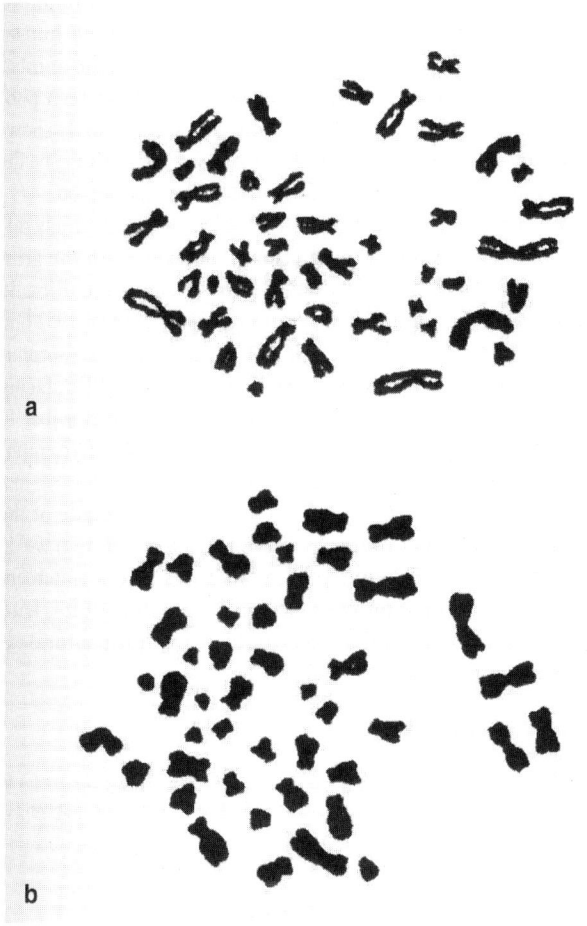

Addendum 2

Ford CE and Hamerton JL (1956). The chromosomes of man. *Nature*, **178**, 1020–1023. Reproduced with permission from Nature Publishing Group.

THE CHROMOSOMES OF MAN

By Dr. C. E. FORD and J. L. HAMERTON

Medical Research Council Radiobiological Research Unit, Atomic Energy Research Establishment, Harwell, Berkshire

ACCURATE knowledge of the number of chromosomes and of their behaviour in mitosis and meiosis has illuminated the genetic analysis of many species. The outstanding example, of course, is *Drosophila melanogaster*[1]. *Zea mais*[2] and *Oenothera lamarckiana*[3] are notable examples from the plant kingdom. Chromosome observations have been of some value even in the fungus *Neurospora crassa*[4], and recent work[5] has given promise that the genetics of the house mouse may benefit in the same way. There is therefore good reason for supposing that the genetic study of man himself may be advanced by reliable information about his chromosomes. The unique features of the human genetic milieu—the alterations in breeding structure[6], the changing selection pressures[7], and the possible influence upon mutation-rates of the environmental changes of the past two or three centuries—are all in great need of study and measurement in order to be able to estimate their effect upon the genetic structure of future human populations, and it is possible that a thorough examination of the chromosomes in a range of human groups would make a useful contribution. The rapid and economical technical methods now available would enable relatively large numbers of individuals to be examined and should make a new approach to human cytogenetics both possible and rewarding. Nor is this the only field in which chromosome observations may be of value to human biology. The rapid pace at which knowledge of the chromosomes in tumours of experimental animals[8] is advancing suggests that the chromosomes may have an increased part to play in the investigation of human neoplastic conditions; and they may also be able to assist in the causal analysis of infertility, at least in the male[9].

The first attempt to determine the number of chromosomes in human cells would appear to have

No. 4541   November 10, 1956   NATURE   1021

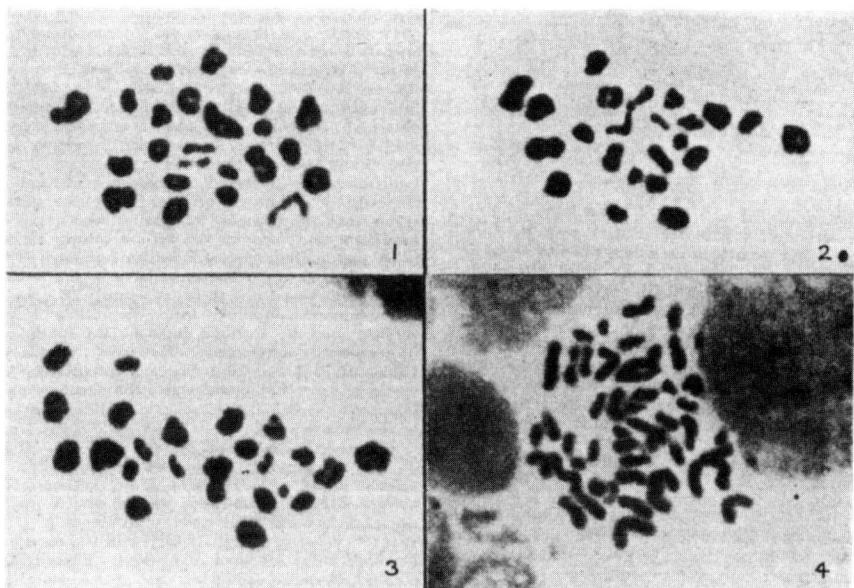

Fig. 1. First spermatocyte of patient 1 at diakinesis, 23 bivalents
Fig. 2. First spermatocyte of patient 3 at diakinesis, 23 bivalents
Fig. 3. First spermatocyte of patient 2, 22 bivalents plus univalent X- and Y-chromosomes
Fig. 4. Spermatogonial metaphase of patient 1, 46 chromosomes
All × 2,800

been made by Hansemann, who in 1891 reported three cells from 'normal human tissues' with 18, 24, and more than 40 chromosomes, respectively[10]. From then until the appearance of de Winiwarter's classical paper[11] in 1912, diploid numbers ranging from 16 to 36 were reported, the balance of opinion being in favour of 24. De Winiwarter claimed that there were 47 chromosomes at metaphase in spermatogonia, and 23 autosomal bivalents plus an unpaired X in primary spermatocytes. Painter[12] in 1921 reported the presence of a small Y chromosome in males and was the first to assert that the correct diploid number was 48 in both sexes[13]. In the following two decades most authors[14] supported Painter's position; but de Winiwarter and his associates[15] adhered to the opinion that there was only a single sex-chromosome at meiosis in the male and 47 chromosomes in spermatogonia. Koller's account[16] of the behaviour of the sex-chromosomes in spermatocyte meiosis brought this period to an end just before the outbreak of the Second World War. From then on, the value of $2n = 48$ in both male and female remained unchallenged for nearly twenty years, and it seemed that the chromosome number of man had finally been established. However, in a very recent paper[17], Tjio and Levan report consistent counts of $2n = 46$ in cultures of lung tissue taken from four aborted embryos, and refer to further counts of $2n = 46$ obtained by Hansen–Melander, Melander and Kullander in preparations of liver, also from aborted embryos. The regular loss of two chromosomes during development of the organs or growth of the cultures would seem to be most unlikely, and the implication was clear that the generally accepted

figure of $2n = 48$ might, after all, have been in error.

A re-examination of spermatocyte and spermatogonial chromosomes was obviously desirable, and we have been fortunate in securing testis tissue from three males, aged forty-seven, fifty-three, and sixty-three years respectively. The material was obtained at the Churchill Hospital, Oxford, from fresh operative specimens at the moment of removal from the body. The tubules were teased out in hypotonic fluid, allowed to remain in the fluid for 30 min. at room temperature, fixed in acetic alcohol and stained by the Feulgen procedure. Squash preparations were made in the usual way. Many of them contained one or more small clusters of spermatocytes in the most suitable stages for counting, namely, diakinesis and first metaphase. Obviously broken cells with scattered chromosomes, and cells in which the chromosomes were clumped or badly fixed, were rejected. Counts were made on all the remainder and are summarized in Table 1. It will be seen that the great majority contained 23 bivalents, and that our results therefore complement those of Tjio and Levan. Examples of these cells are shown in Figs. 1 and 2. In a few cells at late diakinesis and first metaphase, the X- and Y-chromosomes were unpaired and usually well separated from each other, so that 24 bodies were counted (Fig. 3). Precocious disjunction (already reported by Koller[16] in sectioned material) would seem to be the most likely explanation of the non-association of X and Y in these exceptional cells, although some cases may have been due to the rupture of the terminal connexion between them during the making of the preparations. The

Table 1. COUNTS OF NUMBERS OF BODIES PRESENT IN FIRST SPERMATOCYTES AT STAGES FROM LATE DIPLOTENE TO METAPHASE

Patient	Age	First spermatocytes containing:		
		22 bivalents or fewer	23 bivalents	22 bivalents + $X + Y$
1	63	9	81	11
2	53	3	39	11
3	47	2	29	3
Total		14	149	25

few cells recorded as containing fewer than 23 bivalents, although apparently intact, may have been damaged and we attach no significance to them.

Relatively few spermatogonia were observed in mitosis and, in contrast to the spermatocytes, most of them were broken and their chromosomes scattered. We have noticed similar fragility of spermatogonia in several other species of mammals. Nevertheless, a few clear counts of 46 chromosomes were obtained in apparently intact cells, one of which is shown in Fig. 4. It is noteworthy that the centric constrictions are sometimes greatly elongated in from one to four chromosomes.

In the great majority of first spermatocytes examined, the sex chromosomes were associated terminally, the attachment varying from a relatively long thin thread to a condition in which only a slight narrowing indicated where X ended and Y began. All these associations could be interpreted as terminal chiasmata, although Sachs[18] denies that true chiasmata occur in the sex-bivalent of man. The few cells in which the sex-chromosomes were present as univalents have been mentioned. In one cell only the X- and Y-chromosomes appeared to be associated in a different manner. The structure of this bivalent was not resolved with certainty; but its appearance strongly suggested the presence of a sub-terminal chiasma, or possibly that it was a 'symmetrical' sex-bivalent as described by Koller[16]. This single observation can scarcely be construed as support for the possibility of crossing over between X and Y: on the other hand, the observations as a whole are certainly not inconsistent with the occurrence of partial sex-linkage[19].

Koller[16] figured diplotene bivalents with many chiasmata. This observation we confirm. The largest autosomal bivalent frequently has five chiasmata; and some of the others may have four. In well-fixed cells the successive loops can be clearly seen in planes at right angles to each other, as in the classical plant and orthopteran species. In cells where the bivalents are rather crowded, it would not be difficult to mistake a terminal loop for an additional small bivalent. The numbers of chiasmata were counted in some of the clearest cells at stages from late diplotene to mid-diakinesis, and the results are given in Table 2.

Since the hypothesis that each cytological chiasma represents a single genetic cross-over was first put forward, evidence in its favour has been steadily accumulated, although its general validity has been questioned[20]. Assuming it to be true in man, the counts of chiasmata can be used for making an estimate of the total genetic length of the human chromosome set, on the basis that one chiasma is equivalent to 50 centimorgans. The over-all mean number of chiasmata per cell is 55·9 and the estimate of genetic length is, therefore, 27·9 morgans. However, should there have been any reduction of chiasmata by terminalization, the observed frequencies will have been somewhat less than the original frequencies, with the result that the estimate of genetic length will be a minimum one. So far as we are aware, the only other mammalian species in which estimates of total genetic length of chromosomes have been obtained is the house mouse, and it is satisfactory to note that the estimates obtained from Slizynski's chiasma counts[21] and by Carter's linkage-data method are in reasonable agreement. The two methods yield values of 19·2 and 16·2 morgans respectively[22]. It would appear that, genetically, the chromosomes of man are at least half as long again as the chromosomes of the mouse.

The reservation should be made that the estimate of 27·9 morgans strictly applies to middle-aged and elderly males only. However, in the mammalian species studied genetically, recombination fractions do not in general differ much with age or sex. As a rule the values are slightly higher in the female sex, although there are some exceptions. There is therefore no reason for supposing that genetic lengths will be greatly different in the two human sexes, and until direct estimates have been obtained at oogenesis the value of 27·9 morgans derived from chiasma counts at male meiosis may be regarded as an approximate minimum estimate for the female also.

Two comments in connexion with the high frequency of occurrence of chiasmata are relevant: it reflects the rarity with which autosomal linkages have been detected in man; and it implies a genetic system in which there is a rapid re-assortment of the genotypic variability, which, according to Darlington[23], is the mark of a plastic species able to adapt itself readily to diverse environmental circumstances.

Returning to the question of the chromosome number of man, how should the discrepancy between the recent counts and the older ones be explained? Although the occurrence of numerical chromosomal polymorphism has now been established within populations of one mammalian species[24] and may exist in a second[25], it is most unlikely that the occurrence of a similar situation in the human species could provide the whole answer—the probability of drawing by chance from such a polymorphic population seven successive individuals (four of Tjio and Levan, three recorded here) with 46 somatic chromosomes after an even larger series all with 47 or 48 would be altogether too low. Nevertheless, the rare occurrence of individuals with the latter numbers is not excluded. The alternative is to assume a persistent error, and this we consider to be the more likely explanation. (It is of interest to note that in his preliminary communication on the subject, Painter was uncertain whether the diploid number was 46 or 48.) Three features of the behaviour of human chromosomes which have already been mentioned may have contributed to erroneous counting. In somatic cells and spermatogonia, one or more of the exceptionally long centric constrictions seen in some chromosomes may have been overlooked, with the result that each arm was counted as a separate

Table 2. COUNTS OF CHIASMATA AT STAGES FROM LATE DIPLOTENE TO MID-DIAKINESIS

Patient	Age	Cells counted	Chiasmata per cell	
			Range	Mean
1	63	11	50–62	54·4
2	53	6	52–62	57·1
3	47	6	50–63	57·1
Total		23	50–63	55·9

chromosome; and in spermatocytes, counts of 24 bodies could have arisen as a consequence of precocious disjunction of X- and Y-chromosomes, or from mistaking an end loop of one of the larger elements for a separate bivalent. In view of the much closer packing of the chromosomes in the cells, these difficulties would have been much more serious with the older sectioning technique. Undoubtedly the adoption of the squash method[26] and ancillary treatments for dispersing the chromosomes within the cells[27] is bringing about a great change in mammalian chromosomal cytology, and it is to this technical improvement that the rectification of the error—if such it be—must primarily be attributed. The crux lies no longer in the microscope but in the preparative technique. The weary hours of toil which the pioneers must have spent at the microscope is reflected in de Winiwarter's *cri de coeur*, "J'ai perdu un temps énorme à répéter des numérations fatigantes et j'avoue aussi, très fastidieuses". The wonder is that there is so little to alter.

We acknowledge gratefully the help and interest of Mr. G. E. Maloney, who provided us with the material upon which the study was based. The efficient technical assistance of Mr. G. Breckon is also acknowledged.

[1] Dobzhansky, Th., "Genetics and the Origin of Species" (2nd edit., Columbia Univ. Press, New York, 1941). White, M. J. D., "Animal Cytology and Evolution" (2nd edit., University Press, Cambridge, 1954).
[2] Rhoades, M. M., and McClintock, B., *Bot. Rev.*, 1, 292 (1935).
[3] Darlington, C. D., *J. Genet.*, 24, 405 (1931). Catcheside, D. G., *J. Genet.*, 33, 1 (1936).
[4] McClintock, B., *Amer. J. Bot.*, 32, 671 (1945).
[5] Slizynski, B. M., *J. Genet.*, 50, 507 (1952); *J. Genet.* (in the press).
[6] Darlington, C. D., *Nature*, 152, 315 (1943).
[7] Muller, H. J., *Amer. J. Human Genet.*, 2, 111 (1950).
[8] Levan, A., *Ann. N.Y. Acad. Sci.*, 63, 774 (1956). Makino, S., and Kano, K., *J. Nat. Cancer Inst.*, 15, 1165 (1955). Sachs, L., and Gallily, R., *J. Nat. Cancer Inst.*, 16, 803 (1956).
[9] Harvey, C., in "Studies on Fertility, 1955", 8 (Blackwell, Oxford, 1955).
[10] Hansemann, D., *Virch. Arch.*, 123, 356 (1891).
[11] Winiwarter, H. de, *Arch. Biol.*, 27, 91 (1912).
[12] Painter, T. S., *Science*, 53, 503 (1921).
[13] Painter, T. S., *J. Exp. Zool.*, 37, 291 (1923).
[14] Evans, H. M., and Swezy, O., *Mem. U. California*, 9, 1 (1929). Kemp, T., *Z. Mikr. Forsch.*, 16, 1 (1929). Andres, A. H., and Vogal, I. I., *Z. Zellforsch.*, 24 (1936). King, R. L., and Beams, H. W., *Anat. Rec.*, 65, 165 (1936).
[15] Oguma, K., and Kihara, H., *Arch. Biol.*, 33 (1923). Winiwarter, H. de, and Oguma, K., *Arch. Biol.*, 40, 541 (1930). Oguma, K., *J. Morph.*, 61, 59 (1937).
[16] Koller, P. C., *Proc. Roy. Soc. Edin.*, B, 57, 197 (1937).
[17] Tjio, J. H., and Levan, A., *Hereditas*, 42, 1 (1956).
[18] Sachs, L., *Ann. Eugenics*, 18, 255 (1954).
[19] Haldane, J. B. S., *Ann. Eugenics*, 7, 28 (1936); *Proc. Roy. Soc.*, B, 135, 147 (1948).
[20] Cooper, K. W., *J. Morph.*, 84, 81 (1949).
[21] Slizynski, B. M., *J. Genetics*, 53, 597 (1955).
[22] Carter, T. C., *J. Genetics*, 53, 21 (1955).
[23] Darlington, C. D., "The Evolution of Genetic Systems" (Camb. Univ. Press, 1939).
[24] Sharman, G. B., *Nature*, 177, 941 (1956). Sharman, G. B., Ford, C. E., and Hamerton, J. L. (unpublished results).
[25] Wahrman, J., and Zahavi, A., *Nature*, 175, 600 (1955).
[26] Sachs, L., *Heredity*, 6, 357 (1952).
[27] Makino, S., and Nishimura, I., *Stain Tech.*, 27, 1 (1952). Ford, C. E., and Hamerton, J. L., *Stain Tech.* (in the press).

CHAPTER 3

1959: the chromosome basis of Down's syndrome

THE DOOR TO MODERN HUMAN cytogenetics had now been opened, so from our present perspective of clinical cytogenetics, one might have expected that Tjio and Levan's 1956 paper, well publicised internationally, would have led to an immediate flood of clinical applications and discoveries, but in fact it took three years until this began. This should give us no surprise, though, when we look at the situation existing at the time. Most cytogenetics research, in both America and Europe, was being done in basic research laboratories, for whom the human species had previously had no special significance and indeed was unpromising material by comparison with insect and other species with larger and fewer chromosomes. These workers had no incentive to pursue human cytogenetics further and also had no medical links. Those few laboratories studying human chromosomes, like those of Levan in Lund and TC Hsu in Texas, were interested mainly in cancer, which already clearly showed a wide, but at that time apparently inconstant, range of chromosome abnormalities in tumour cells.

Also, particularly in America, the developing human and medical genetics departments in universities were headed by people without particular experience or interest in cytogenetics, which had a low academic profile by comparison with mathematical, population, or even clinical genetics.

Cytogenetics as a diagnostic tool in clinical pathology laboratories was essentially non-existent at this time, while the idea that it might become valuable in the study of constitutional disorders had not crossed the minds of any but a very few. The nearest approach was the use of the sex chromatin body, already known since 1949, as described in Chapter 1, and which was beginning to be applied through buccal smears to such conditions as Turner and Klinefelter syndromes. On top of all of this, tissue culture of adult material was still difficult, so chromosome analysis usually required material from surgical tissue samples or from testicular or bone-marrow biopsies, invasive and unpleasant procedures unlikely to be readily available for a systematic series of patients – or indeed from normal volunteers as controls. The use of small and essentially painless superficial skin biopsies for fibroblast culture by Harnden[1] and others (see Chapter 7) was only just beginning by 1959.

Despite these negative aspects, Tjio and Levan's discovery and its rapid confirmation by groups in other countries and using other tissues, caught the interest of human geneticists, who realised that there was now the possibility of determining whether particular genetic disorders were caused by a major chromosome abnormality. The most important potential candidate was Down's syndrome, then universally known as 'Mongolism'.

Since the term 'Mongolism' has long since been abandoned, but occurs throughout the papers reproduced and quoted here, it is perhaps worth recalling the remark of TC Hsu on the subject.

Some people may wonder why I, a Chinese, keep using the somewhat racist term Mongol or Mongolism instead of the more neutral term Down's syndrome... perhaps the story I heard about Mongolism would make Orientals less resentful: a Westerner once asked a Japanese paediatrician how he would describe the features of Mongols. Said the doctor, 'look like Caucasians'.[2]

The possibility that Down's syndrome might be due to a chromosomal abnormality had been raised as long ago as 1932 by the Dutch ophthalmologist and geneticist Waardenburg, as can be seen below:

The stereotyped recurrence of a whole group of symptoms among the Mongoloids offers an especially fascinating problem. I would like to suggest to the cytologists that they examine whether it may be possible that we are dealing with a human example of a certain chromosome aberration. Why should it not occur occasionally in humans, and why would it not be possible that – unless it is lethal – it would cause a radical anomaly of constitution? Somebody should examine in Mongolism whether possibly a 'chromosomal deficiency' or 'nondisjunction' – or the opposite, 'chromosomal duplication' – is involved .... My hypothesis at least has the advantage of being testable. It would also explain the possible influence of maternal age.[3]

This was also suggested by the American geneticist Davenport in the same year. Davenport's reputation has been clouded by his extreme eugenic views, but his note on Down's syndrome is worth recording here since he actually sent material to Painter for chromosome analysis.

Since we now know that aberrations in the chromosomal complex are responsible for irregularities of development in both plants and animals, it is reasonable to look for them in man also. Herein may lie the cause of some profound defects that are clearly familial, but the method of whose inheritance is not easily revealed. It would seem that, if anywhere, we should find such chromosomal irregularities in the group of the feeble minded. Some years ago I was able to assist Painter to get some perfectly fresh testicular material of a mongoloid dwarf. But, Painter tells me, this material revealed no obvious chromosomal irregularities. However, this negative result should not discourage us from continuing the search for possible chromosomal irregularities in genetically complex defects. Such chromosomal irregularities have, indeed, been found in cancer cells; they are, consequently, not foreign to human tissues, nor probably, to human gametes.[4]

A chromosomal basis was also seriously considered by Lionel Penrose,[5] at the Galton Laboratory, London, who by the 1950s had become the acknowledged world authority on Down's syndrome, and indeed in human genetics overall. Penrose favoured the possibility of triploidy rather than trisomy and in 1952 asked his colleague Ursula Mittwoch, working with him at the Galton, but not specifically on chromosomes and with most inadequate facilities, to examine a testicular sample from a Down's patient. She tentatively concluded that the likely number was 47 or 48 and, in the light of the then universal view that the normal chromosome number was 48, concluded that there was probably no gross chromosome abnormality in Down's syndrome and that it was certainly not the result of triploidy.[6]

Yet again we can see the harm that can be caused by erroneous preconceived ideas that have become entrenched in the literature. Although Ursula Mittwoch herself, in an interview over 50 years later, was fully aware of the technical limitations of her study, Penrose seems to have firmly turned against a chromosomal basis for Down's syndrome, since in his otherwise masterly 1954 review on the disorder, this aspect receives only a brief mention:[7]

Gross cytological abnormality, such as polyploidy or fragmentation of chromosomes, is excluded since the germ cells of a typical case were found to be normal in appearance (Mittwoch 1952).

In Penrose's chapter on the disorder in the 1955 book, *La Progénèse*, edited by Raymond Turpin, chromosomes are not even mentioned.[8]

While Mittwoch's study might have been helped by the use of a control, normal volunteers for testicular biopsy would have been hard to find. As TC Hsu states laconically in his book[3] (writing, significantly, from Texas):

there were only two ways to obtain human testicular material: waiting outside the operating rooms and waiting by the gallows.

LIONEL PENROSE (1898–1974)

Penrose was born into a Quaker family in London, serving during World War I in the Friends Ambulance Service as a pacifist, and then training in medicine. Specialising in psychiatry he took a post as research officer under the Medical Research Council at the Royal Eastern Counties Institute, Colchester, starting his lifelong work on the basis of mental handicap, especially Down's syndrome, and leading to the famous 1938 'Colchester Study', the first systematic scientific analysis in the field.

In 1946 Penrose was appointed to the Galton Chair of Eugenics at University College London (a fervent anti-eugenicist, he changed its title and that of its associated journal to 'Human Genetics'). Under Penrose the Galton Laboratory became the main centre in the UK, and the world, for human genetics research, especially theoretically based work, attracting many workers from other countries who would become leaders in medical genetics, though interestingly neither clinical genetics nor cytogenetics were strongly represented at the Galton.

Penrose's ability as a thinker, and his powerful though extremely unassuming personality, gave an exceptional cohesion and loyalty to those who had worked with him. His strongly principled approach was a key factor in re-establishing human genetics after its abuse in many countries in the years leading up to and during World War II. Although a collection of memories and an informal life of Penrose have been written,[26,27] and his records archived,[28] there remains a need for a critical scientific biography of this key figure in human genetics. (Photograph courtesy of Professor Shirley Hodgson.)

Nevertheless, chromosome studies in Down's syndrome were beginning in a number of centres across Europe, including Edinburgh (Patricia Jacobs), Guy's Hospital, London (Paul Polani) in conjunction with Harwell (David Harnden and Charles Ford) and at Uppsala, Sweden (Marco Fraccaro and Jan Lindsten in the Institute of Jan Böök). The most definitive study, however, and the first to be published, was that in Paris, involving Drs Jérôme Lejeune, Marthe Gautier and the head of the paediatric department at Hôpital Trousseau, Raymond Turpin. Turpin had already been engaged on a broad study of the condition for a number of years[9] and had collaborated with Penrose on aspects such as fingerprint patterns (dermatoglyphics). A recent article gives valuable information on the background to Turpin's work on Down's syndrome and on his role in the trisomy 21 discovery.[10]

Lejeune's name is now the one most associated with the discovery of trisomy 21 in Down's syndrome, understandably so in the light of his lifelong involvement in the chromosome field, but the cytogenetic study was largely based on the skills and work of his colleague and co-author Marthe Gautier, now the only surviving member of the three authors of the original papers, whom I was able to interview in 2004 and 2005. Her name is less well known than it deserves to be since she returned soon after the discovery to the field of clinical paediatrics and paediatric cardiology, but her work is covered in a valuable historical review of early human cytogenetics in France and elsewhere, by Simone Gilgenkrantz.[11] Marthe Gautier was already an experienced clinical paediatrician and investigator, who had recently returned from a research fellowship in Philadelphia, where she had learned cell culture techniques, before joining the group of Professor Raymond Turpin, with whom Lejeune was already working. She set up a laboratory at Hôpital Trousseau in Paris, under the most primitive conditions, bartering chicken's eggs with the nearby Institut Pasteur and keeping a cockerel in the hospital courtyard as a supply of low calcium serum, as essential tissue culture ingredients. (Several other early investigators also recount stories about their cockerels!)

The microscopy, undertaken by both Marthe Gautier and Jérôme Lejeune, was also basic (Marthe Gautier had to purchase a microscope herself). It is in fact a remarkable testimony to their skill and perseverance that preparations were as clear as they were (Figure 3-1). The tissue used was mainly *fascia lata* (taken with parental permission) from which fibroblasts were cultured; although skin was also used, in particular from the investigators as controls, this was less successful due to overgrowth of epithelial cells. The results, recorded in the the Hôpital Trousseau laboratory book, are preserved today at Hôpital Necker (Figure 3-2), to which Lejeune's laboratory later moved, and it can be seen that the first trisomic individual was discovered in May 1958. Since the study had been planned from the beginning as a series, numbers steadily increased so that Lejeune was able to take a considerable amount of data with him when he attended the August 1958 International Human Genetics Congress in Montreal. Turpin had apparently instructed him to say nothing about the discovery at the congress, but Lejeune did present the findings informally at a subsequent seminar at the Department of Genetics, McGill University. Dr Clarke Fraser, who organised the seminar, remembers (personal communication 2005) that:

Jerome was quite diffident in his presentation. He showed photos (no karyotypes of course); there certainly seemed to be an extra blob, present in all 7 'mongols' and no controls, which he suggested might be an extra chromosome. I thought it was pretty

Fig. 3-1 Original preparation of trisomy 21 chromosomes by Dr Marthe Gautier, 1958. Colchicine was not used in this preparation. (Courtesy of Dr Marthe Gautier.)

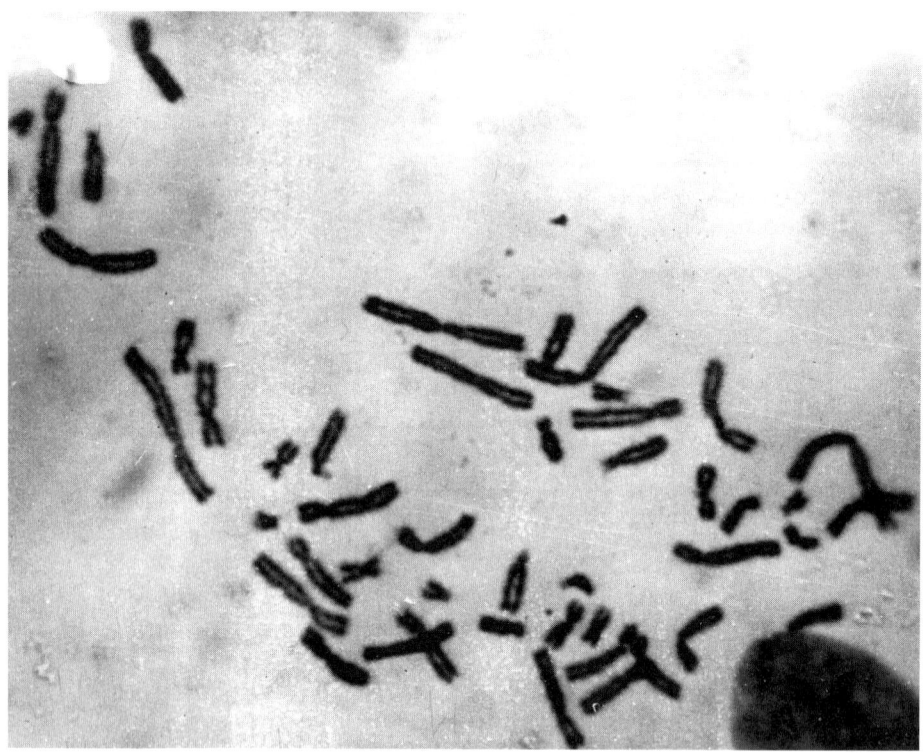

Fig. 3-2 Entries in the Hôpital Trousseau lab book showing dates of initial findings in the Paris Down's syndrome study. I am most grateful to Dr Marguerite Prieur for access to these archives and for allowing reproduction of the material shown here.

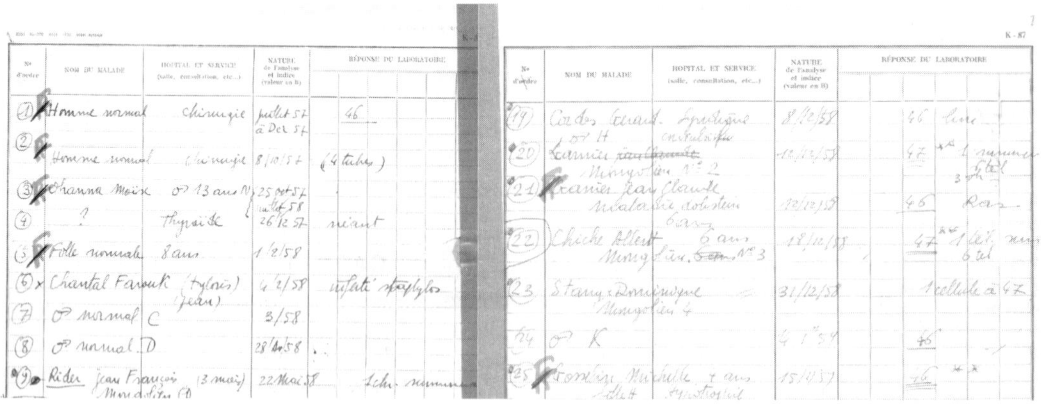

MARTHE GAUTIER

Born into a farming family, she was encouraged by her mother, at considerable sacrifice, to study medicine, beginning her studies in Paris in 1942 and joining her older sister, who was later killed in the Nazi retreat from Paris. An outstanding student, despite the severe difficulties then for women in medicine, she entered paediatrics, working with Professor Robert Debré, who arranged for her to go to Harvard, where she learned tissue culture techniques in relation to rheumatoid arthritis.

On her return, she found that she had been 'forgotten' and the post promised her given to another, which led to her joining Professor Raymond Turpin and setting up a tissue culture lab for cytogenetic studies in Down's syndrome, as described in the text. Relationships with Lejeune were not easy, however, particularly for a strong minded and independent woman, and she later returned to clinical work, making a distinguished career in paediatric cardiology. Now retired, she divides her time between Paris and her country home, where she paints on porcelain. (Photograph courtesy of Dr Marthe Gautier.)

JÉRÔME LEJEUNE (1926–1994)

Lejeune began working on Down's syndrome with Raymond Turpin in 1952, studying a range of possible aetiological factors before concentrating on the cytogenetic study with Marthe Gautier. After the discovery of trisomy 21 he extended his cytogenetic research to a wide range of abnormalities, his laboratory becoming the focus for the development of clinical cytogenetics across France. Lejeune's high international profile and forceful personality ensured that Paris became and remained a major centre for cytogenetic research (see also Chapter 8), but his opposition to prenatal diagnosis and Catholic orthodoxy led to him distancing himself from the mainstream medical genetics community in France and internationally during the later part of his life, as described vividly in the paper of Simone Gilgenkrantz.[11] (Photograph courtesy of Fondation Jérôme Lejeune.)

exciting, but in general the reaction was sceptical.

John Hamerton (personal communication 2005) was also at this seminar and recalls that it produced a 'buzz' of excitement.

Lejeune was insistent on his return that the work must be published rapidly, despite Turpin's hesitation. The first published paper in *Comptes Rendus* of the Academy of Sciences, 26th January 1959,[12] devoted only a single paragraph, given below, to Down's syndrome.

Chez trois garçons mongoliens, le nombre chromosomique trouvé est de 47 sur différentes preparations, pour chacun des trois individus. Dans les trois cas le diagnostic de sexe a été rendue impossible du fait de la présence d'un trés petit chromosome supplementaire qu'on ne peut différencier des cinq petits éléments presque telocentriques, normalement rencontre chez l'Homme.

In three Mongol boys, the chromosome number was found to be 47 in different preparations, for each of the three individuals. In the three cases the diagnosis of sex has been made impossible by the presence of a very small additional chromosome that one cannot distinguish from the five small, almost telocentric elements, normally seen in the male.

The second paper, in March 1959, still brief and without photomicrographs, gave details of nine patients[13] and is given in translation at the end of this chapter; an additional small acrocentric chromosome was again observed in all patients. A more detailed account with photomicrographs (see Figure 3-3) appeared in the 14th April,

RAYMOND TURPIN (1895–1988)

Profoundly affected by his work during World War I as a young medical officer, Turpin entered paediatrics and initially worked mainly on tuberculosis and other infectious diseases. His growing interest in developmental disorders led to his programme of research on Down's syndrome, joined first by Jérôme Lejeune and then by Marthe Gautier. His book, *La Progénèse*, covers much of what would today be considered as human and medical genetics, and he set up the Institut de Progénèse in 1958, later moving across Paris to become Head of Paediatrics at Hôpital Necker. It is of interest that almost all the founders of human and medical genetics in France were practising paediatricians, an influence that remains strong today. (Photograph courtesy of Dr Marie-Hélène Couturier-Turpin.)

Fig. 3-3 Trisomy 21 (a) chromosome preparation and (b) karyotype from Lejeune et al., Figures 5 and 6.[14] (Courtesy of the Académie Nationale de Médecine, Paris.)

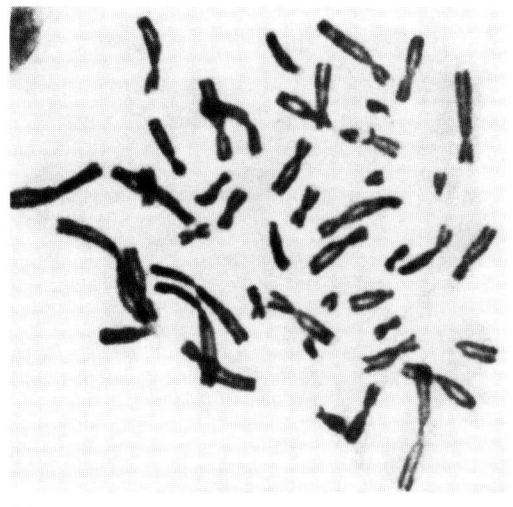

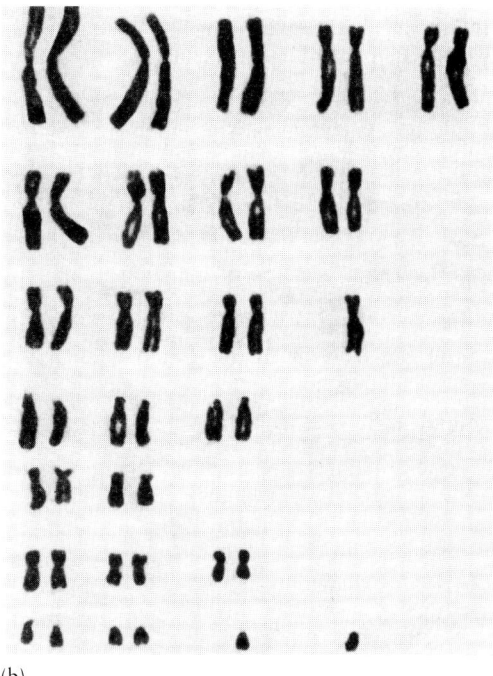

(a)

(b)

1959 issue of the *Bulletin of the National Academy of Medicine*.[14]

Fortunately for the group, publication in *Comptes Rendus* seems to have been a remarkably rapid procedure, even by present day electronic standards. The story current in Paris today (possibly apocryphal) is that the best way to ensure prompt publication was to speak to a relevant academician after Sunday Mass and give him (it was always a man) the brief manuscript, which could then, if suitable, be read at the Academy's weekly meeting on Monday and printed and published in a week or so! However, even more important was the intensity with which Gautier and Lejeune carried out the project, both realising that other groups were working on the same lines.

Lejeune and colleagues were probably wise to study a considerable series rather than an isolated case, otherwise the significance of this small extra chromosome might have been disputed, given technical limitations, but its occurrence in all the patients was convincing and to an extent unexpected, although Turpin's expert clinical experience would have minimised the possibility of heterogeneity. The existence of a sub-group due to translocation would only emerge the following year.

The fact that the chromosomal basis of Down's syndrome first emerged from France, a country with little previous experience or technological basis in cytogenetics, surprised the international community at the time and still seems surprising today; given the very different, forceful and potentially conflicting temperaments of the key individuals involved (to speak of the 'Paris group' would not be really correct), it is remarkable that the work reached its initial goal without major difficulties. Yet, to the great credit of those involved, it did and as a result founded a lasting and outstanding tradition of human cytogenetics in Paris that was to bring further discoveries in later years, as described in Chapter 8.

Marthe Gautier left the field of genetics in 1963 to return to clinical work; in interviews with the author in 2004 and 2005 she emphasised that this has always been for her the most important aspect of her distinguished career. None the less, it is also clear that she retains great pride in her early laboratory work and the discovery of trisomy 21. Raymond Turpin continued his involvement with cytogenetic developments, his unit, including Lejeune's lab, moving across Paris from Hôpital Trousseau to Hôpital Necker when he was appointed as Head of Paediatrics there. Jérôme Lejeune devoted himself to further chromosome research and became the international standard-bearer for French cytogenetics until the advent of prenatal diagnosis a decade later. His vehement opposition to this, related to his strongly orthodox Roman Catholic beliefs, then resulted in a major division within French human cytogenetics which many workers feel seriously weakened this field in France.[11]

A number of people in France consider that Lejeune's outspoken views cost him the Nobel prize. The Nobel committee's records remain closed for 50 years, but I myself doubt that this factor was significant. Lejeune had undoubtedly antagonised colleagues in Paris (notably Maurice Lamy, Head of Medical Genetics at Hôpital Necker), but with the prize limited to three individuals, it would have been difficult (and wrong) to ignore the vital contributions of Marthe Gautier and Raymond Turpin, or the simultaneous discovery of the sex chromosome anomalies by other groups. It is a pity, though, that human cytogenetics should not have been recognised in some way by a Nobel award, particularly given the multiple awards to those in molecular biology; this would have helped to emphasise the outstanding value of the early cytogenetics work and to correct the perception that molecular genetics was a 'superior' field scientifically and intellectually, a perception strongly promoted by a number of molecular biologists themselves.

Studies elsewhere soon confirmed that Down's syndrome was indeed due to trisomy. The series of Patricia Jacobs and her colleagues in Edinburgh[15] was probably the most extensive and most advanced by the time of the Paris workers' publication; this owed much to the close involvement of an experienced clinician, John Strong, who liaised with the local mental handicap service to obtain samples. The work of the Edinburgh group will be taken up in more detail in the next chapter, but Patricia Jacobs' recollections of the Down's study are relevant, especially the role of Penrose in helping to confirm the clinical status of the patients.

PJ. *We went, this was in 1958, and got bone marrow from I think it was 30 or 40, a not inconsiderable number of patients who the institution told us were Down's. We looked at the chromosomes and most but by no means all, appeared to have an extra chromosome – 47 in all. This was true whether they were males or females, but all of them didn't. Out of those 40, about 6 didn't. We went back to the institution and we said 'most of them seem to have an extra chromosome but they don't all. What can this mean? Have they all got Down's syndrome?' The institution assured us that they did. So I thought, I wonder if this is true. We decided the best thing we could do was to get an expert on Down's to come up and look at them blind. So we asked Penrose to come up. Lionel Penrose came all the way from London and we all went out to this institution, I was there too, just to look, and these 40 patients all paraded in front of Lionel and it was an education to me. He was my hero ever since that day, to see how he dealt with these patients and the way they responded to him. It was great and he went right through and he told us which ones were Down's, without knowing this. And which ones were Down's plus something else, and which ones didn't have Down's. He said it so nicely that even the people in the institution were willing to accept it. And of course he got it completely right and all the ones he said were Down's had the extra chromosome, and the ones that we hadn't found the extra chromosome in, he had absolutely no doubt, didn't have Down's.*

PSH. *So there wasn't a translocation Down's in that group?*

PJ. *None. Not a single translocation Down's in that group. They were all trisomy, which was fair enough. We might have expected to find one but we didn't.*

Marco Fraccaro, originally from Pavia (and later to return there), along with Jan Lindsten, then a medical student, both working at this time in the Human Genetics Institute at Uppsala, Sweden, with Professor Jan Böök, were also able rapidly to confirm the trisomic basis of Down's syndrome;[16] they used the techniques of Tjio and Levan, and material from skin fibroblasts and bone marrow from three patients. It should be noted that the Uppsala unit, founded on the work of the outstanding geneticist Gunnar Dahlberg,[17] who had died relatively young, and subsequently headed by Jan Böök, was, along with the London Galton laboratory, the only comprehensive human genetics institute at that time, and thus attracted young workers from across the world, including Fraccaro from Italy and Kurt Hirschhorn from America. Hirschhorn and

MARCO FRACCARO (born 1925)

Marco Fraccaro studied medicine at the University of Pavia, specialising in pathology. Recognising the future importance of genetics in medicine from the work of Buzzati-Traverso, also in Pavia, he spent a year under Penrose at the Galton Laboratory, London, before taking a post in 1956 at the newly formed Institute of Human Genetics in Uppsala, Sweden, under Jan Böök. There he established a cytogenetics laboratory and was involved in some of the first descriptions of chromosome abnormalities, including trisomy 21 and sex chromosome anomalies. After two years in the Oxford unit of Stevenson he returned to Pavia in 1962, where he founded human cytogenetics in Italy and developed a flourishing wider department of medical genetics over the subsequent decades. (Photograph courtesy of Professor Marco Fraccaro.)

Orlando J Miller, both subsequently pioneers of human cytogenetics in the United States (see Chapter 8), also spent time at the Galton Laboratory. Malcolm Ferguson-Smith, who had developed cytogenetic studies in Glasgow (see Chapter 4), was at this time at Johns Hopkins School of Medicine, Baltimore, emphasising the close transatlantic links.

At Charles Ford's Harwell unit, not hospital-based, but with clinical links, notably with Paul Polani, already established primarily for the study of possible sex chromosome disorders (see Chapter 4), David Harnden was now developing the less invasive skin fibroblast culture technique,[18] well-suited to long distance collaborations, and again Penrose was enlisted in the study.[19] The following is an extract from an interview with Professor David Harnden on 18th March 2004.

I was in the library one day reading about Down's syndrome and how 50% of the offspring of female Down's patients also had Down's syndrome and I thought what's going on? I knew about Pat's discovery on Klinefelter's and I thought, Down's syndrome is a chromosome abnormality. I got in touch with Penrose and he got a biopsy of a patient with Down's syndrome, sent it to me and I had it in my incubator growing up when I heard that both Lejeune and Jacobs had discovered the 47 chromosomes in Down's syndrome. But that was very interesting because it turned out to be a patient who had 48 chromosomes. He was both a Klinefelter and a Down's. It was the first double aneuploidy, and the other interesting thing about that, when we got this worked up, it was Penrose who drove it. I did the work and when we wrote

JAN LINDSTEN (born 1935)

Jan Lindsten entered Uppsala Medical School in 1955, but interrupted his studies to undertake cytogenetics research with Marco Fraccaro in 1957. Although he was involved in the early Down's syndrome studies, his special interest became Turner syndrome, the subject of his PhD thesis. After completing his medical studies, he was appointed Professor first in Aarhus, Denmark, then in Stockholm, where he built up a comprehensive medical genetics department. Later he became Medical Director of the Karolinska Hospital, Stockholm and was also Secretary of the Swedish Nobel Prize committee. (Photograph courtesy of Professor Jan Lindsten.)

the paper up it turned out that the authorship was Harnden, Miller and Penrose; I said to Lionel, 'You know really you're the boss. You should be first author.' He said, 'No, all my papers are in alphabetical order', so it was Harnden, Miller and Penrose – that was OJ Miller from Columbia University. So that was my contribution to Down's syndrome. A bit too late to get the glory!

It seems most unlikely that Penrose did not appreciate the special features of this patient (a brief note had already appeared in *The Lancet*[20]), so David Harnden's approach provided a rapid and relatively non-invasive way of achieving cytogenetic analysis after Penrose, who was the world authority on the condition, sadly had largely missed out on the primary discovery of the trisomic basis of Down's syndrome. It was not until 1960 that cytogenetic analysis was systematically established at Penrose's Galton Laboratory by Joy Delhanty.

In 1959 there was no generally agreed system of nomenclature for human chromosomes (often a vexed topic, considered more fully in Chapter 7), and distinguishing between the group of small acrocentric chromosomes, including the Y chromosome, was far from easy. The Paris group considered that the extra chromosome they had found in Down's syndrome belonged with the larger pair they termed Vh, later known as 21; they discuss this fully in their detailed (April 1959) paper,[14] from which Figure 3-3 is taken. In fact the pair involved is the smaller of the two, as shown later by Levan and by Ferguson-Smith[21] and so the disorder logically should have been reclassified as 'trisomy 22', but by that time the terms 'Down's syndrome' (or Mongolism) and 'trisomy 21' had become so inter-related that change was impossible. An example of the Paris nomenclature is given in Chapter 7.

Translocation Down's syndrome

Neither the Paris nor the Edinburgh series of Down's patients included any of the now familiar sub-group resulting from translocation rather than trisomy. The existence of a distinct sub-group born to younger mothers had already been recognised, notably by Penrose,[5] and when trisomy 21 was established, this sub-group was clearly of importance to study separately. Penrose himself was the obvious person to initiate such a study, but his lack of cytogenetic laboratory facilities gave Paul Polani, also in London (Guy's Hospital) and already closely linked with Charles Ford's Harwell unit through their work on Turner syndrome (see Chapter 4), the first opportunity to demonstrate translocation Down's syndrome. Polani had been seconded to the World Health Organisation in the summer of 1958 as part of an international study on pregnancy loss, and was travelling between Copenhagen, London and America, shortly after setting up the sex chromosome collaboration with Charles Ford. Not yet aware of the Paris work, he reassessed the possible chromosomal basis of Down's syndrome. The following is an extract from Polani's unpublished memoir, 'The Year is 1959' (I am most grateful to Professor Paul Polani for permission to quote from this document).

The WHO Europe HQ was in Copenhagen where I settled for a while and was given free access to the University Library, and an especially welcome chance to discuss sex chromosomes and sex determination with M. Westergaard. It was here that, browsing, re-reading classic human chromosome papers and the more recent ones, and especially Ford and Hamerton's meiotic in vivo confirmation (n = 23) of Tjio and Levan's finding, it was here that Ursula Mittwoch's paper on a male mongol's meiotic chromosome number (n = 24) came to mind. Had she miscounted or was there something wrong with the Down syndrome chromosome number?

Immediately he heard from Charles Ford that an abnormality had been found in Turner syndrome, he initiated the Down's study, specifically including patients born to younger mothers.

Sex apart, urgently we were to study chromosomally patients with Down syndrome. We felt that trisomy of a small autosome was a possibility. We also decided that, at some second time – and depending on our first findings in Down syndrome – we should consider studying Down syndrome patients born to younger women. Such cases were much rarer than those born from older mothers but sibship clustering of affection had been noted in the young-mother group, which suggested that they might represent a distinct subgroup of Down syndrome.

No sooner had I arrived at the Bethesda Campus than Ford telephoned to tell me that Lejeune had examined the chromosomes of nine Down syndrome subjects. Their chromosome number was 47: they carried a small extra chromosome. In short they turned out to be trisomic for the chromosome later defined as No. 21: Ursula Mittwoch had counted correctly!

It was likely that Lejeune's patients were children of older women. So we went ahead with the study of Down syndrome from younger women, as planned, and thereby identified translocation (centric fusion) Down syndrome, 'the mongol with 46 chromosomes' (1960). The condition, a 'hidden' trisomy of the Down syndrome chromosome translocated to another chromosome, was genetically transmissible through asymptomatic translocation carriers, formally with 45 chromosomes. I reported this at the December meeting in New York of the American Neurological Society at which Lejeune reviewed his work. Later, the centric fusion detection allowed us to calculate the mutation rate for that structural chromosome anomaly; while, at the applied level, it was the impetus for the prenatal detection of the Down anomaly by amniotic cell culture.

Polani and Ford's findings were published in The Lancet in April 1960;[22] nomenclature was still fluid at this point so the additional translocated chromosome is actually referred to in their paper as 22. They did not analyse parental samples, but pointed out the likely heritable nature of the translocation; later that year two further papers, both from London, appeared in The Lancet which clearly showed the familial nature of translocation Down's syndrome. Penrose, with Joy Delhanty who had by now established cytogenetics at the Galton Laboratory, was able to show from skin biopsies that the healthy mother, grandmother and a sib of a translocation Down's patient all showed the translocation and had 45 chromosomes[23] (Figure 3-4), while Polani's group, now joined by John Hamerton and collaborating with Cedric Carter at the Hospital for Sick Children in London, reported a comparable three generation family.[24] This illustrated for the first time how cytogenetic analysis could be useful for genetic counselling by identifying a high-risk group among the Down's patients born to younger mothers and was itself a landmark for the practical application of human cytogenetics.

Fig. 3-4 Translocation Down's syndrome. Chromosome preparation and karyotype of balanced translocation carrier (from Penrose et al., 1960)[23]. (Reproduced from *The Lancet* with permission from Elsevier.)

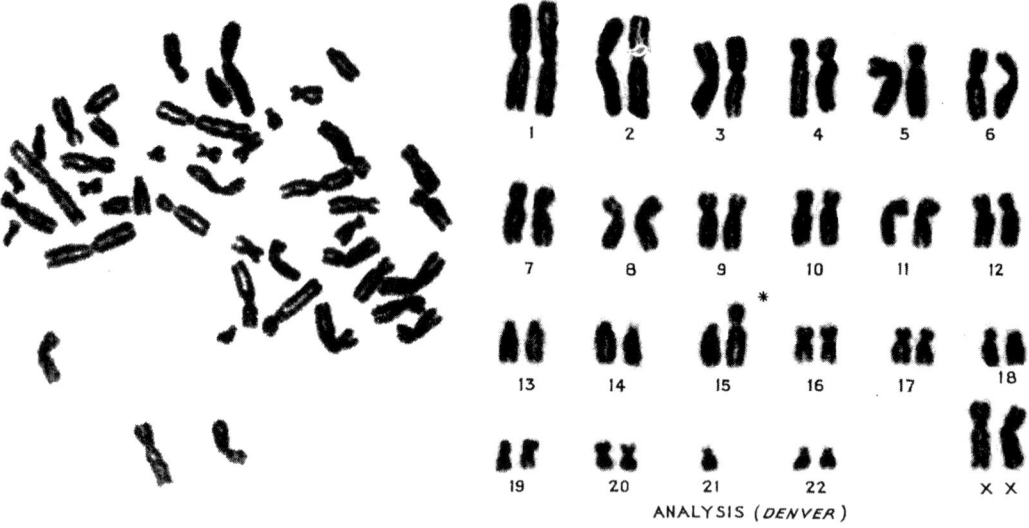

Mosaicism and Down's syndrome

Although experienced clinicians were able to identify most Down's patients clearly on clinical grounds, as seen with Penrose and the Edinburgh series, a small number of confusing cases still remained. Cytogenetic analysis was not only able to identify these as mosaics, but to establish mosaicism as a general phenomenon in genetic disorders, where previously it had merely been a possibility. Mosaic Down's syndrome, following on from mosaicism in the sex chromosome disorders, was again first shown by Ford's Harwell Laboratory, with John Edwards (see Chapter 5), then working in nearby Oxford, as the clinician involved in this study.[25] In their case, a two-year-old girl, the facial features and palm prints suggested Down's syndrome, but her developmental progress was normal.

Skin biopsies showed both a normal and a trisomic cell line, confirming the clinical value of both the cytogenetic analysis and and the use of skin fibroblasts. Interestingly, analysis of a blood sample (this was now late 1960) was normal, reflecting the fact that while skin fibroblasts would soon be superseded by blood for most diagnostic purposes, they would remain important for the detection of mosaicism.

Down's syndrome had for many years

been the paradigm for congenital disorders of likely genetic origin, but until 1959 no laboratory-based genetic investigations had yielded any significant understanding over that given by clinical and epidemiological studies. Now, with its trisomic basis firmly established, more detailed studies of its origin and the underlying aetiological factors could begin. Even more importantly, there was good reason for a wider range of clinicians and pathologists to become interested and to create the demand that would ensure that chromosome investigations were not only a part of basic cytogenetic research, but also of diagnostic investigation – in essence, to create the discipline that within a short space of time would become identifiable as clinical cytogenetics.

References

1. Harnden DG (1960). A human skin culture technique used for cytological examination. *Br. J. Exper. Pathol.* **41**, 31–37.
2. Hsu TC (1979). *Human and Mammalian Cytogenetics. An Historical Perspective.* New York, Springer.
3. Waardenburg PJ (1932). *Das Menschliche Auge und seine Erbenlangen.* Den Haag, Nijhoff, pp. 47–48. (Translation by authors in Vogel F and Motulsky A (1986). *Human Genetics: Problems and Approaches.* New York, Springer.)
4. Davenport CB (1932). Mendelism in man. Procedings of the Sixth International Congress of Genetics, Ithaca, New York, 1932, Vol. 1, pp. 135–140. (I am greatly indebted to Dr James Crow for pointing out and finding this reference.)
5. Penrose LS (1938). *A Clinical and Genetic Study of 1280 Cases of Mental Defect.* London, HMSO, pp. 36–37.
6. Mittwoch U (1952). The chromosome complement in a Mongolian imbecile. *Ann. Eugenics* **17**, 37.
7. Penrose LS (1954). Observations on the aetiology of Mongolism. *Lancet* **2**, 505–509.
8. Penrose LS (1955). La Part réelle de l'hérédité dans les oligophrénies. In: Turpin R (ed.), *La Progénèse.* Paris, Masson, pp. 224–223.
9. Turpin R (1954). Effect of maternal age on frequency of malformations. *Bull. Acad. Natl Med.* **138**, 433–435.
10. Couturier-Turpin M-H (2005). La découverte de la trisomie 21. *La Revue du Practicien* **55**, 1385–1389.
11. Gilgenkrantz S and Rivera EM (2003). The history of cytogenetics. Portraits of some pioneers. *Ann. Génétique* **46**, 433–442.
12. Lejeune J, Gautier M and Turpin R (1959). Les chromosomes humains en culture de tissus. *C. R. Acad. Sci.* **248**, 602–603.
13. Lejeune J, Gautier M and Turpin R (1959). Étude des chromosomes somatiques de neuf enfants mongoliens. *C. R. Acad. Sci.* **248**, 1721–1722.
14. Lejeune J, Turpin R and Gautier M (1959). Le mongolisme, maladie chromosomique. *Bull. Acad. Natl Méd.* **143**, 256–265.
15. Jacobs PA, Baikie AG, Court Brown WM and Strong JA (1959). The somatic chromosomes in mongolism. *Lancet* **1**, 710.
16. Böök JA, Fraccaro M and Lindsten J (1959). Cytogenetical observations in mongolism. *Acta Paediatr.* **48**, 453–468.
17. Böök JA (1957). Gunnar Dahlberg; in memorium. *Acta Genet. Stat. Med.* **6**, I-III.
18. Harnden DG (1996). Early studies on human chromosomes. *BioEssays* **18**, 163–168.
19. Harnden DG, Miller OJ and Penrose LS (1960). The Klinefeltermongolism type of double aneuploidy. *Ann. Hum. Genet.* **24**, 165–169.
20. Ford CE, Jones KW, Miller OJ, Mittwoch U, Penrose LS, Ridler M and Shapiro A (1959). The chromosomes in a patient showing both mongolism and the Klinefelter syndrome. *Lancet* **1**, 709–710.
21. Ferguson-Smith MA (1993). From chromosome number to chromosome map: the contribution of human cytogenetics to genome mapping. *Chromosomes Today* **11**, 3–19.
22. Polani PE, Briggs JH, Ford CE, Clarke CM and Berg JM (1960). A Mongol girl with 46 chromosomes. *Lancet* **1**, 721–724.
23. Penrose LS, Ellis JR and Delhanty JDA (1960). Chromosomal translocations in mongolism and in normal relatives. *Lancet* **2**, 409–410.
24. Carter CO, Hamerton JL, Polani PE, Gunalp A and Weller SDV (1960). Chromosome translocation as a cause of familial mongolism. *Lancet* **2**, 678–680.
25. Clarke CM, Edwards JH and Smallpiece V (1961). Trisomy/normal mosaicism, in an intelligent child with some Mongoloid characters. *Lancet* **1**, 1028–1030.

26. Smith M. *Lionel Sharples Penrose: a biography*. (Privately published; undated. ISBN 0 9535780 0 3.)
27. Centre for Human Genetics, University College London (1998). Penrose: Pioneer in Human Genetics. Report on a symposium held to celebrate the centenary of the birth of Lionel Penrose. Centre for Human Genetics, UCL.
28. University College, London (1979). A list of the papers and correspondence of Lionel Sharples Penrose (1898–1972).

Addendum

Lejeune J, Gautier M, and Turpin R (1959). Étude des chromosomes somatiques de neuf enfants. *Comptes Rendues de l'Académie des Sciences*, **248**, 1721–1722.
English translation reproduced from *Papers on Human Genetics* (1963) by SH Boyer (Ed.), with permission from Prentice Hall.

Étude des chromosomes somatiques de neuf enfants mongoliens
(Study of the Somatic Chromosomes of Nine Mongoloid Idiot Children)

JEROME LEJEUNE
MARTHE GAUTIER
RAYMOND TURPIN

The culture of fibroblasts from nine Mongoloid children reveals the presence of 47 chromosomes, the supernumerary chromosome being a small telocentric one. The hypothesis of chromosome determination of Mongolism is considered.

The study of mitosis of fibroblasts in culture from nine Mongoloid children recently (1) permitted us to establish with regularity the presence of 47 chromosomes. The observations made in these nine cases (five boys and four girls) are recorded in the table below.

The number of cells counted in each case may seem relatively small. This is due to the fact that only the pictures that claim a minimum of interpretation have been retained in this table.

NUMBER OF CELLS EXAMINED IN EACH CASE

		Diploid cells						Tetraploid cells		Total
	Number of chromosomes	"Doubtful" cells			"Perfect" cells			"Perfect" cells		
		46	47	48	46	47	48	—	94	—
Boys	Mg 1	6	10	2	—	11	—	—	1	—	30
	Mg 2	—	2	1	—	9	—	—	—	—	12
	Mg 3	—	1	1	—	7	—	—	2	—	11
	Mg 4	—	3	—	—	1	—	—	—	—	4
	Mg 5*	—	—	—	—	8	—	—	—	—	8
Girls	Mg A	1	6	1	—	5	—	—	—	—	13
	Mg B	1	2	—	—	8	—	—	—	—	11
	Mg C	1	2	1	—	4	—	—	—	—	8
	Mg D	1	1	2	—	4	—	—	—	—	8
		10	27	8		57			3		105

* This child is the product of a twin pregnancy. His normal twin, examined at the same time, possesses 46 chromosomes, of which 5 are small telocentric ones.

The apparent variation in the chromosome number in the "doubtful" cells, that is to say, cells in which each chromosome cannot be noted individually with certainty, has been pointed out by several authors (2). It does not seem to us that this phenomenon represents a cytologic reality, but merely reflects the difficulties of a delicate technique.

It therefore seems logical to prefer a small number of absolutely certain counts ("perfect" cells in the table) to a mass of doubtful observations, the statistical variance of which rests solely on the lack of precision of the observations.

Analysis of the chromosome set of the "perfect" cells reveals the presence in Mongolian boys of 6 small telocentric chromosomes (instead of 5 in the normal man) and 5 small telocentric ones in Mongoloid girls (instead of 4 in the normal woman).

The "perfect" cells of non-Mongoloid individuals never present these characteristics (1) and therefore it seems legitimate to conclude that there exists in Mongoloids a small supernumerary telocentric chromosome, accounting for the abnormal figure of 47.

DISCUSSION

To explain the sum total of these observations, the hypothesis of non-disjunction of a pair of small telocentric chromosomes at the time of meiosis can be considered. It is known that in Drosophila non-disjunction is greatly influenced by maternal aging, such a mechanism accounting for the increase in frequency of Mongolism as a function of the advanced age of the mother.

It is, however, not possible to say that the supernumerary small telocentric chromosome is indeed a normal chromosome and at the present time the possibility cannot be discarded that a fragment resulting from another type of aberration is involved.

REFERENCES

(1) J. Lejeune, M. Gautier and R. Turpin. 1959. Comptes rendus. **248:** 602.

(2) P. A. Jacobs and J. A. Strong. Nature. 1959. **183:** 302-303.

CHAPTER 4

The sex chromosomes

ALTHOUGH THE CHROMOSOMAL BASIS of Down's syndrome has been covered first in this book, the search for abnormalities in the sex chromosomes shows in fact a longer and more focused background; both came to fruition in 1959, which Paul Polani has termed 'the wonderful year' of human cytogenetics.

To understand the sequence of events we must return briefly to the beginnings of genetics, when the mechanisms of sex determination and role of the sex chromosomes were first worked out over a 20-year period between 1891 and 1910. This work was done almost entirely on insect chromosomes, but it established the conceptual framework so strongly that the later mammalian and human studies were all initially interpreted in terms of this theoretical basis.

EB Wilson's masterly 1911 review, 'The sex chromosomes',[1] gives a synthesis of the different studies of the previous 20 years, while Carlson's recent book[2] gives a clear sequential account of how our knowledge evolved, well illustrated by inclusion of original figures from the different papers. The monograph of Mittwoch[3] is also a valuable source of information on work up to the mid-1960s.

The first step in our understanding was Henking's description in 1891[4] of an unusual body beneath the nuclear membrane that appeared to migrate as a chromosome during meiosis but which remained condensed. He did not pursue the work further, but his 'X element' (so called because its nature was unknown) later gave the name to the X chromosome once its chromosomal nature was fully established by McClung,[5] Paulmier[6] and Sutton[7] in 1899–1900. McClung proposed that this chromosome was sex determining,[8] but it soon became clear that patterns of sex determination and sex chromosomes varied considerably between different species. In 1905 both Stevens[9] and Wilson[10] found that in some insects (including *Drosophila*) there was an unequal pair of chromosomes in males (XY), the females being XX, and by 1910 Wilson[11] was able to set out the general principle of sex determination, with one sex (usually the male) being 'heterogametic' (XY or XO) and the other (usually the female) being 'homogametic' (XX).

For the first attempts to study the situation in humans, Painter's work, already referred to in Chapters 1 and 2, provided the foundations. Painter's 1921 paper[12] and its successor[13] established that in the opossum a Y chromosome was present in male meiosis, and his extensive 1923 study of human chromosomes showed the same to be true for man.[13] Figure 4-1 shows one of Painter's original illustrations of XY pairing; he makes it clear in his paper, though, that morphologically the Y chromosome could not be distinguished with certainty from other small acrocentric chromosomes, a limitation that would remain true for almost another 50 years, until the advent of fluorescence techniques. The same, of course, also applied for the larger X chromosome except in its condensed form.

None of this work seemed at the time to be of relevance to human disorders of sexual differentiation or to medicine generally, but this changed immediately with the discovery of the sex chromatin by Murray Barr and Mike Bertram in 1949[14], described in Chapter 1. It had soon been shown that the presence or absence of sex chromatin was a reliable indicator of sex in humans, while Keith Moore's development of the buccal smear technique[15] made it a potentially valuable investigation in the study of human intersex and related disorders, some of which would later prove to be due to abnormalities of the sex chromosomes. Two conditions, Turner and Klinefelter syndromes, were to be especially central to this work.

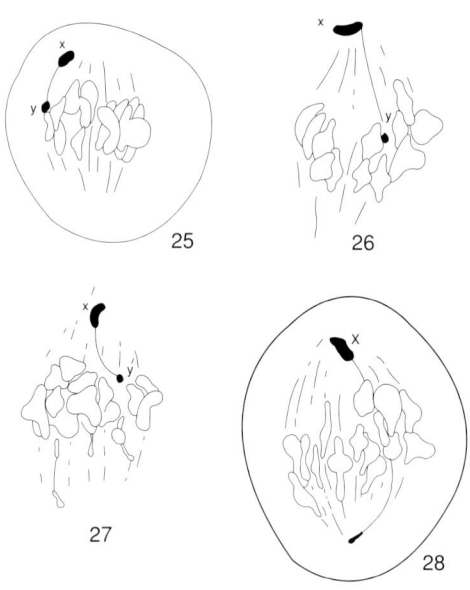

Fig. 4-1 Human X and Y chromosomes during spermatogenesis (from Painter, 1923).[13]

Turner syndrome had been recognised as a clinical disorder a number of years previously,[16] the main features being ovarian dysgenesis and a range of other physical abnormalities. Prominent among those investigating its basis was Paul Polani (see box), based at Guy's Hospital, London, who had been led to the condition via a study of congenital heart disease. Polani's unpublished biographical essay 'The year is 1959' sets the stage for how the work developed.

The year is 1959, the year of the 'human cytogenetics explosion'. For me it is the year when work over the last ten years was coming to fruition. Research on

> **PAUL POLANI (born 1914)**
>
>
>
> Paul Polani was born in Trieste and studied biology and medicine first in Siena, then in Pisa at the Scuola Normale Superiore. Arriving to do postgraduate work in Britain at the outbreak of World War II, he initially worked as a ship's surgeon in the Merchant Navy, but when Italy entered the war he was interned on the Isle of Man. Later, he was posted to the Evelina Children's Hospital in London's East End, where he was the only resident medical doctor until the end of the war.
>
> In 1948 he moved to nearby Guy's Hospital to undertake research, first on kernicterus, then on congenital heart disease, at the same time learning genetics from Lionel Penrose at the Galton Laboratory. This work led to his discoveries on the chromosomal basis of Turner syndrome and translocation Down's syndrome, and subsequently, joined by John Hamerton, to wider cytogenetic research. In 1960 the UK Spastics Society decided to found a major institute at Guy's Hospital to undertake research on cerebral palsy and related developmental disorders; Paul Polani was appointed as director, building it into the largest and most comprehensive medical genetics research institute in the country and developing also a full medical genetics service alongside it as part of the National Health Service.
>
> Now aged over 90, Paul Polani remains actively involved and interested in the field of medical genetics, of which he was one of the original pioneers. A short, informal biography is included in the 1982 Festschrift volume for his retirement,[33] which also contains a series of contributions from his colleagues. (Photograph courtesy of Professor Paul Polani.)

congenital cardiac anomalies and my study of human genetics – e.g. 'moonlighting' at the Galton with Penrose – were now falling into place. However, it was my findings and conjectures on Turner syndrome, arising out of the congenital heart work, and more indirectly on Klinefelter syndrome, that were important and had confirmed cytogenetically my ideas on abnormality of chromosomes, more especially of sex chromosome complements and their bearing on sex determination in man. The relatively high frequency in Turner syndrome females of coarctation of the aorta, a male cardiac anomaly, had suggested to me that these females might be sex-reversed males, and I found that their cell nuclei tested chromatin negative (1954). The male frequency of red–green colour blindness reported in The Lancet (1956) had confirmed the presence of a single X chromosome, but now I thought that Turner syndrome patients might be XO (45,X) females, rather than XY sex-reversed males: which suggested that, if XO, sex determination in man (and perhaps in mammals generally) could not be as in Drosophila *(where XOs are male), then the accepted formula also for humans.*

(I am greatly indebted to Professor Paul Polani for access to and permission to quote from this unpublished autobiographical memoir.)

Polani describes in an interview in late 2003 how he was led to the sex chromatin and subsequent work.

POLANI. *So in dealing with the material which Maurice Campbell [cardiologist] had accumulated, I was left with a bunch of about, I forget exactly, let's say 15 coarctations, which I didn't know what to do with because there were too few; but we had an idea then already that it was a male malformation because out of those 15 fewer than 5 were females. Four females I think, and of the 4 females, 2 or 3 had ovarian agenesis and they were thought to be Turner syndrome, webbing of the neck, small and so on; they weren't children of course, they were already young adults. So the question that came to mind first was whether this was a mimic by nature or the other way round... let's say a mimic by nature of the Jost experiments of the castration of rabbits where if you castrate a rabbit it turns out a female phenotype....*
I wondered whether these were in fact originally females who had lost their ovaries, because in those days we thought they didn't have any ovaries: in fact, ovarian agenesis was their classification. And so how to test for it was a bit of a problem, but we decided just to use the Barr body which had just come off the production line, as it were, and I did that work with Hunter and Lennox to whom I sent specimens, which I had already examined myself but didn't trust my judgement.

PSH. *Did you have a lab at Guy's then?*

POLANI. *Yes, I had; there was an interdepartmental laboratory, a new idea for Guy's which was extremely handy. So I collected, I think 3 or 4 bits of skin, sent them to Lennox and said would you please tell me what you think about nuclear sexing of these bits of skin; and time went by and nothing happened. So one day I rang him up and said 'What have you done? What have you made out of those specimens?' He said 'They are not very interesting, just ordinary males'. Exactly what I wanted! But the second thought, as soon as he said that, I started thinking really more seriously and I said 'right; well what is the evidence that they are males?' We knew that we were testing only for the X, through Barr body testing. And I just wondered whether they might be XO and in order to confirm that they had only one X chromosome at any rate, irrespective of whether the other one was missing or was a Y, I thought about using some sort of genetic marker on the X. Haemophilia was one which would be too rare to use, but* colour blindness seemed to be OK; and when I discussed it with Penrose he said 'Oh yes, colour blindness would be fine, because it is about 7% for males and 0.5% for females, so you would be alright if you have enough patients'. So I got enough patients: I had been helped by Sir Robert Platt, who then was Dr Platt, and by other people who gave me patients, Peter Bishop particularly at Guy's. So I got my 25 patients and out of those 25, three were colour blind. And there was a nice story there, a personal story attached to it because I turned out to be colour blind myself, which I didn't know!

Interestingly, Lionel Penrose, who had been Polani's mentor in genetics over the preceding decade, was most reluctant at this point to accept the possibility that Turner syndrome was due to an XO chromosome basis, and thus that human sex determination must differ from the classical model provided by Drosophila, as the interview makes clear:

PSH. *At what point did you manage to convince Penrose that human sex determination was different from* Drosophila?

POLANI. *Penrose would not have it! Penrose would not have it, I have to say. He was annoyed with me! He said, 'Where do you get this stupid idea?' And I said, 'Well, yes Professor Penrose, but see, the figures would suggest that there is something'. 'Yes but you know, I mean all sorts of . . .'. Well anyway, when I sent my paper into* The Lancet *in 1956 I had the audacity not only of suggesting that they might be XO sex, but also writing that, if indeed they were XO sex, they would be unlike what happens in* Drosophila. *Here were females of XO sex where this was officially male. So perhaps sex determination in man was not correctly interpreted in those days. And* The Lancet *wouldn't have this bit!*

PSH. *Do you think* The Lancet *might have got Penrose's views?*

POLANI. *I don't know. They certainly got some views and I know that they were quite determined for me to alter this bit, because the editor then rang up Philip Evans whose friend he was, and he knew that I was working for Philip Evans, and said, 'No we can't have that sort of thing. Get him to modify it. Take it all out.' And I said 'No. I'm not going take out the XO sex story'!*

Polani's findings of both absent sex chromatin in Turner syndrome[17] and a male distribution of colour blindness[18] made it clear that human sex determination in general, as well as Turner syndrome itself, could not be adequately explained on the classical *Drosophila* model and pointed urgently to the need for detailed chromosome analysis. Here, the collaboration with Charles Ford at the UK Medical Research Council's Radiobiology Unit at Harwell provided the key link. Charles Ford (see box, Chapter 2) was an expert basic cytogeneticist whose principal interest was radiation-induced chromosome damage in both plants and mammals. Although he had been responsible for developing human bone marrow chromosome analysis (with Patricia Jacobs and Laszlo Lajtha; see also Chapter 6), as well as the confirmation in 1956 (with John Hamerton) of the 46 human chromosome number (see Chapter 2), he was reluctant to be drawn further into human work from his primary interest of mammalian cytogenetics in relation to irradiation. Polani takes up the story in his essay:

In 1955 I sought a direct refutation, or confirmation, of my unorthodox views on the sex chromosomes in Turner's and Klinefelter's, and hence on human sex determination. I first turned to Gordon Thomas, an expert at Guy's on tissue culture. However, though we tried hard and indeed saw and counted chromosomes in cultured somatic cells of both patients and controls, we came to no useful conclusion. Having approached, but without success, P Koller for his excellence with chromosomes, I turned for help to Charles Ford at the MRC Radiobiology Unit, Harwell. I had come to know him in relation to the sex chromatin and colour blindness work, and we became better acquainted during the 1957 Nuclear Sexing Symposium at King's College Hospital. In fact, I had suggested that Ford be invited, given his expertise in cytogenetics generally and his recent work on murine chromosomes.

In the Spring of 1958 I received two preprints of the same paper that was to appear in Nature, *one sent by Ford and the other by L Lajtha, concerning the bone marrow technique for the study of human chromosomes. I immediately sent to Ford bone marrow from patients with Turner and Klinefelter syndromes (and from myself).*

At the end of 1958, while in Copenhagen, I heard from Ford the exciting results of the chromosome work in our two patients, a Turner female with 45 chromosomes, considered XO, and a Klinefelter male with 47 chromosomes, seemingly XXY (he also had a normal cell line with 46 chromosomes, so probably he was a sex-chromosome mosaic). It was for me, if I may quote 'the Eighth Day of Creation', 'one of those rare and exciting moments when observation, surmise and experimental results snap into soul-satisfying harmony'.

The paper of Ford, Polani and colleagues[19] appeared in *The Lancet*, almost at the same time as Lejeune's *et al.* first report on trisomy 21 (see Chapter 3) and that of Jacobs and Strong in *Nature* on Klinefelter syndrome, discussed below. High profile medical and scientific journals were now taking a keen interest in chromosomes, a change from the situation when the human chromosome number was first reported three years before.

Simultaneously to the studies on Turner

syndrome, cytogenetic work was in progress on one of the other main disorders of sexual development, Klinefelter syndrome. Again this had been well delineated as a clinical disorder involving hypogonadism and absent spermatogenesis in 1942,[20] and it also had been found in 1956 to show a sex chromatin body, despite the patient's male phenotype.[21]

One worker closely involved was a young trainee pathologist in Glasgow, Malcolm Ferguson-Smith, who recalls his first work on Klinefelter syndrome in an interview in 2004 with the author, of which a fuller sequence is given in the accompanying recording:

MALCOLM FERGUSON-SMITH (born 1931)

Born and educated in Glasgow, Malcolm Ferguson-Smith trained there in pathology, developing his life-long interest in sex chromosome disorders (see text) before moving in early 1959 to Johns Hopkins Hospital, Baltimore, to develop human cytogenetics there with Victor McKusick. Returning to Glasgow in late 1961, he founded a comprehensive medical genetics unit, being one of the first to develop prenatal diagnosis, while continuing his cytogenetic and sex chromosome research, notably on the mechanisms of XY pairing. Moving to the Chair of Pathology in Cambridge in 1987 and making key contributions to gene mapping and molecular techniques in cytogenetics, he now continues his work on comparative molecular cytogenetics at the Cambridge Veterinary School.

During that time I came under the wing of Dr Bernard Lennox, who was at one time a pathologist at the Hammersmith and had just come, about a year or so before I arrived on the scene, to Glasgow. He was the pathologist who worked with Paul Polani in identifying the sex chromatin bodies or the absence of sex chromatin bodies in Turner's syndrome.

Now, after a little while a young man who was being examined for the Army, a medical test, was found to have small testes. He was referred to see what was the cause of his hypogonadism; he was quite a tall, strapping man, he had a testicular biopsy and I found a tubule full of spermatogenesis. Anyway, in amongst the spermatogenesis I was very excited to find a sperm and lots of spermatocytes [Figure 4-2], *and in the spermatocytes, I was aware that there were things called sex vesicles; I knew what the sex vesicles were, they were where the X-Y bivalent was present.*[22] *So I went along to Bernard Lennox and I said, 'Look, these guys are not sex-reversed females, because this guy has*

Fig. 4-2 Klinefelter testicular biopsy showing (a) active spermatogenesis in one tubule and (b) abnormal spermatozoon with visible tail (from Ferguson-Smith).[22] (Courtesy of Professor Malcolm Ferguson-Smith and *Scottish Medical Journal*.)

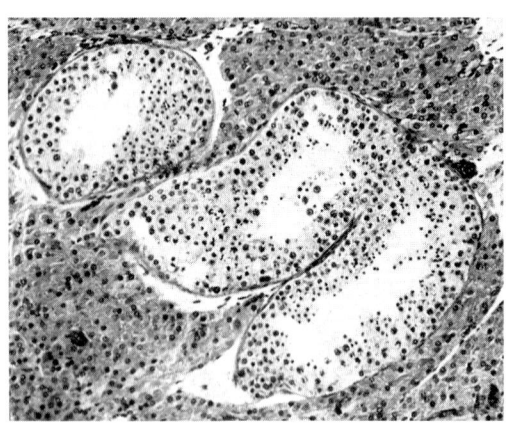

(a)

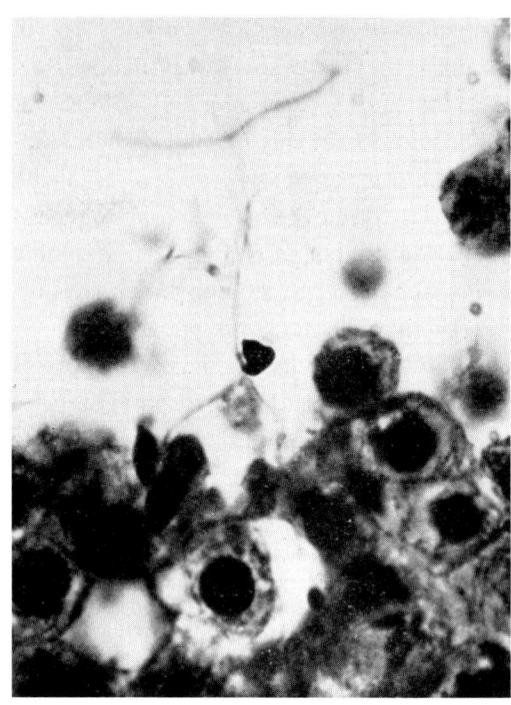

(b)

got a Y chromosome'; we talked about this and said, 'We will have to look at the chromosomes.'

At this point, Ferguson-Smith contacted Charles Ford and learned of the bone marrow technique for chromosomes, but his offer of samples from Klinefelter patients was not taken up by Ford, who included a supposed Klinefelter case from another source in the 1958 bone marrow paper (with Lajtha and Jacobs, see below);[23] this showed an XX karyotype and, with hindsight, was probably a sex-reversed XX male. At the beginning of 1959, therefore, with one of Ford's cases probably clinically misdiagnosed and the other (that referred to above by Polani,[24] and still unpublished at the time) apparently an XXY mosaic, the chromosomal basis of Klinefelters syndrome was puzzling; if Charles Ford had accepted Malcolm Ferguson-Smith's earlier offer any confusion would probably have been avoided. As it turned out, the XXY chromosome constitution was first demonstrated by a third UK group, involving Patricia Jacobs (working at the MRC Clinical Effects of Radiation Unit led by Michael Court Brown in Edinburgh) and an experienced clinical endocrinologist, John Strong, based like the MRC unit at the Western General Hospital in Edinburgh, who was also involved in the Down's study, as mentioned in Chapter 3.

Patricia Jacobs' primary remit was to study chromosome damage in radiation-induced leukaemia and in 1957, immediately after starting work with the Edinburgh MRC unit, she was sent to learn human chromosome techniques based on bone marrow, first with Laszlo Lajtha in Oxford, who had developed bone-marrow culture methods, and then with Charles Ford at the nearby Harwell MRC unit, a key combination of techniques which she was able to link together,[23] as she describes in an interview in 2004:

PJ. *My remit was to look at the chromosomes of people with radiation-induced leukaemia and see if they were any different from the chromosomes of people with leukaemia that was not radiation induced. When I started I went down to Oxford and Harwell to learn two things. From Harwell and from Charles Ford, I was to learn how to look at chromosomes – he didn't really do human chromosomes, he did mouse chromosomes; and from a man called Lazlo Lajtha in Oxford, who was a haematologist who grew bone marrow. As you know, Harwell and Oxford are close together; I went between those two individuals learning what were to be the two aspects of my research project, I suppose. I stayed in Harwell and went between these two labs for about three months.*

PSH. *And you published a paper with Charles Ford on human chromosomes?*

PJ. *And Lazlo Lajtha. That's right, I did that when I was there. But it was basically a technical paper on how you could use bone marrow and Charles' techniques, very beautiful techniques for looking at marrow chromosomes in mice, which was what he was doing at that time, and how you could apply it to humans, using Lazlo's technique, which was for growing bone marrow and nothing to do with chromosomes. So I was the person that linked those two together – a very young and very naive creature.*

The 1958 *Nature* paper of Ford, Jacobs and Lajtha[23] gives the definitive description of the bone marrow technique for chromosome analysis, which was to underlie many of the key advances of the next few years. Returning to Edinburgh at the beginning of 1958 Jacobs was rapidly able to establish the new techniques and also made close scientific and clinical links both within the unit and with other clinical workers, notably with John Strong, which soon proved their value.

I was offered a Klinefelter patient by a very nice, far-sighted I have to say, endocrinologist called John Strong. He said, would you be interested, and I said 'Sure. Why not?'. So he got me marrow; I looked at the Klinefelter and the preparations were really very bad, even though I had practiced. And I thought there were 47 chromosomes and that there were two Xs and a Y; remember we couldn't recognise either of them precisely, not even the Y. But I did think it was XXY and I couldn't believe it. This was not the perceived wisdom of what Klinefelters were in that day and age. There were said to be two kinds of Klinefelters, called chromatin positive and chromatin negative and nobody had clearly made any distinction between them except in their sex chromatin body; everybody assumed that they were sex-reversed females, so everybody expected them to be XX, but I thought I could see 47 chromosomes. And I thought I could see something that was compatible with being a Y, which means there were 5 acrocentrics rather than 4. So I went on holiday and I asked my technician to prepare a tray of slides with the Klinefelter in it and lots of other things in it too. I came back from my holiday and I scored them blind and I thought, well that's funny, because there were two that seemed to have 47 chromosomes, not just one as I had expected. I said to her, 'I've got two that I really think might have 47 chromosomes', and she broke into a big grin because she had put two from the Klinefelter in the tray. So I thought well, that may be true. So that's it.

The interview with Patricia Jacobs illustrates a number of key general points that were highly relevant to the success of their work and whose presence or absence was equally critical to success or failure in other laboratories. She emphasises these also in her 1982 Allan award address,[26] which provides a detailed personal account of her early work. The importance of close links, both physical and intellectual, between laboratory and clinical workers has been mentioned, but the flexibility in those early years of funding bodies such as the UK Medical Research Council must be ranked as a major and far-sighted element, one that has sadly been to a large extent lost today worldwide.

Pat Jacobs' experience of this flexibility was echoed in several of the other interviews. Another striking feature was that many of these key findings were made by people who were very young and at the beginning of their scientific careers, yet they were given a remarkable degree of responsibility and freedom in their work, even if the writing of papers proved something of an ordeal!

We told Michael Court Brown, who saw the significance of it. I didn't. I mean I was 22 years old or something and I knew nothing about how the Y wasn't supposed to do anything, because we were all supposed to be like Drosophila *and that the Y didn't have any effect whatsoever on sex. He seemed to think it was very important and he said, well you had better go and write it up. Now while I had published one paper before, I have to say most of*

PATRICIA JACOBS (born 1934)

Born in London, but living in Scotland from the age of four, Patricia Jacobs studied zoology as an undergraduate at St Andrew's University, doing research there on insect chromosomes under Professor Mick Callan, her first publication being on non-disjunction in the praying mantis, *Mantis religiosa*. After a year's break in America she was offered a post at the newly formed Edinburgh MRC unit in 1957 (see main text and Chapter 6) under Dr Michael Court Brown, applying cytogenetic techniques in relation to leukaemias and radiation. This yielded major discoveries in the field of sex chromosome abnormalities and Down's syndrome as well as in the leukaemias, work that was later extended to population surveys of chromosome abnormalities.

In 1972 she moved to the University of Hawaii, working on the cytogenetics of spontaneous abortions and later on fragile X mental retardation. At this time she married Dr Newton Morton, mathematical geneticist, then also working in Hawaii. In 1985 they moved to New York and after two years to Britain, where she became director of the Wessex Regional Cytogenetics Unit, Salisbury. Here she developed a molecular genetics and cytogenetics diagnostic service as well as continuing further research on the origin of chromosome abnormalities and the basis of fragile X syndrome and related conditions, up to the present time. (Photograph courtesy of Professor Patricia Jacobs.)

MURIEL LEE

Technician with Patricia Jacobs throughout the early work on human chromosomes (see text) she is seen here in the laboratory of the Edinburgh MRC unit. (Photograph courtesy of Mrs Muriel Lee.)

it was written by Mick Callan. It was based on my work on the praying mantis, but Callan wrote it. So I went off and wrote this paper, with great difficulty I have to say, and brought it back and showed it to Michael Court Brown and I have never forgotten that. He tore it in shreds and I stood there nearly in tears, and said 'But it is the first paper I have ever written' and he sort of held it like this [indicating a limp rag!]. *And he said that is exactly what it looks like. So I had another go with his help and that was it. We published that paper and this was the famous 'Jacobs and Strong' in* Nature.[27]

A final element, often inadequately recognised, is the importance of skilled technical staff underpinning the work. The combination of laboratory skills, initiative and enthusiasm for the work are well shown in an interview with Muriel Lee, technician at that time with Patricia Jacobs, but there are many other such 'unsung heroes' whose memories deserve to be documented.

ML. *It was exciting because I was going to night school at that time and I was being taught that there were 48 chromosomes. The human number was 48 and I was going to work the next day and counting 46 and it was just so exciting, because we were right in there at the beginning you know.*

PSH. *Do you remember when about was it that you came up with something that was clearly abnormal?*

ML. *Well the first real abnormal was the Klinefelter, that I can remember; we had got the sample from Professor Strong and we did all the techniques and then Pat actually looked at it and she thought there were 47 chromosomes, but as I said the preps weren't really ideal at that time and so she thought that, but she wasn't absolutely certain. So she asked me, she was going away for a few days and she asked me if I would put in some ones that I knew to be normal and just put this in and mark them all just a, b, c or whatever so she didn't know what they were. So she was looking at them blind, just a tray of slides. I did that and she came out and she was quite excited. She said, 'I think we've actually got two with 47 chromosomes'. And I said, 'Well, that's right because I put in two'. She was quite impressed with that. She has always been quite impressed with that.*

PSH. *So you did that without telling her?*

ML. *Yes. I thought that would be a real test, you know, if she could see it in two slides.*

PSH. *With that patient, were you expecting to see an abnormality or was it something that came out unexpectedly?*

ML. *I wasn't. I'm sure Pat was, but I wasn't really at that time. I did understand that we were looking for chromosome abnormalities, but I really didn't in that specific incident, I don't recall thinking 'this might be it'.*

PSH. *But when she identified it blind, you realised that this was really something?*

ML. *Then I counted and thought this is worth something, you know.*

PSH. *That must have been really very exciting.*

ML. *It really was. It was superb. Superb.*

By the end of 1959 both Turner and Klinefelter syndromes were well characterised chromosomally, but the combination of using sex chromatin as a screening test in selected populations and the feasibility of chromosome analysis on peripheral blood (Chapter 7) soon led to the recognition of further sex chromosome anomalies. First was the XXX female, again discovered by the Edinburgh group in 1959.[25] Patricia Jacobs recalls the discovery:

We studied a lot of patients in mental retardation institutions using sex chromatin. We were very, very fortunate to have a superbly good pathologist called Neil MacLean in the Western General Hospital and he got very interested in doing the sex chromatin for us. And so we did huge surveys of people for sex chromatin abnormalities. And of course by that time we were also interested in triple X women, the first one of which we stumbled upon while looking at patients who had presented in local clinics for menstrual abnormalities. So the first triple X patient actually had that. She had premature ovarian failure and we discovered she had two sex chromatin bodies; that was thanks to Neil MacLean's very astute observation. That was the first patient seen to have two sex chromatin bodies; up till then it was thought the sex chromatin body represented two X chromosomes stuck together. But now you had two sex chromatin bodies, not three, and it was realised that it must be all but one.

By 1960 this group and others had been able to identify a considerable range of different X chromosome abnormalities, some mosaic, and with a wide range of phenotypic features, including Klinefelter patients with more than two X chromosomes, who showed significant mental handicap.[28,29]

The rare males with XXYY chromosomes were also detected by being sex chromatin positive[30,31] and it was the finding in a survey done elsewhere by Casey and colleagues, and at the time unpublished, that these appeared to be over-represented in a sex chromatin survey of high-security hospital patients that led the Edinburgh workers to the possibility that the additional Y chromosome might be responsible and to a detailed chromosome survey to test this hypothesis. This led in turn to the discovery by full chromosome analysis of the more frequent XYY syndrome.[32] Patricia Jacobs here refers initially to this other study, then to her own:

PJ. *A very significant proportion were XXYY. A far increased proportion than you would get in ordinary institutions for the retarded, and I said to him [Casey]: 'That's far too many XXYYs. I am sure it is. I can only remember seeing one in all the patients we've done', and he said 'No, it's just chance'. I thought, I don't think it is. I went back and I looked up our data to be sure what I was saying was right and I found my one and only XXYY was also in under conditions of high security, but in an ordinary institution for the retarded. So I thought, that's very funny isn't it? Maybe the Y is affecting their behaviour. Well that's quite likely isn't it? If you stop and think about it, 98% or some such number of the prison population are males. So you can't say the Y has got nothing to do with behaviour, because why would you get this astounding figure, which is incontrovertible.*

PSH. The main risk factor has always been being male.

PJ. *Exactly; so I went back and I told Michael Court Brown. He said 'Well that's very interesting. Let's see if we can go and look at special hospitals'. Only one in Scotland and we got permission to do this from the Scottish Office; we went back and we looked at these male inmates of which there were several hundred, I think three or four hundred, I can't remember, and 15 females were in that same institution. That was the ratio there. And they try to tell me the Y has got nothing to do with it. Absurd! We were not allowed to see the patients, for reasons I simply don't know; the medical service for these patients was done by the local GP. He was a lovely man, and he would be in one room with the patients, while we were next door with our bottles. We would give him one and he would bring us back the bottle and a tiny piece of paper that had a coded number on it, the patient's date of birth and, for no reason I can think of, the patient's height. We didn't want the patient's height. We never asked for it, but it was there.*

We took these back and we looked at them in the lab. We only got about 10 a week. Week two, and I have never forgotten this, my technician who had been with me right from the beginning, Muriel, she flung open my door and said 'Well, here we go, we've either got an XYY or a 6 foot 4 inches Down's patient!' I thought that was marvellous because, remember, we had a hypothesis that we would find these people there. So I kept very calm and I said 'Muriel, the first one doesn't count, it's the second one that counts. This could be chance.' And the next week we found another one.

The early studies of human sex chromosomes described in this chapter were soon taken up and extended by other workers, including Jan Lindsten in Sweden, Albert de la Chapelle in Finland and Kurt Hirschhorn in New York. They were not only important for their medical and diagnostic applications, but they provided the foundations for a wealth of later research, mentioned briefly in Chapter 8, on the basic mechanisms and evolution of sex determination and differentiation, on X chromosome inactivation and X chromosome biology generally, and on the mapping and conservation of X linked genes. Almost all of this fundamental work can be directly traced back to these beginnings of 50 years ago.

References

1. Wilson EB (1911). The sex chromosomes. *Mikrosk. Anat. Entwicklungsmech.* 77, 249–271.
2. Carlson EA (2004). The sex chromosomes. In:

Mendel's Legacy: The origin of classical genetics. Cold Spring Harbor, CSHL Press, pp. 79–98.
3. Mittwoch U (1967). *Sex Chromosomes.* New York, Academic Press.
4. Henking H (1891). Über Spermatogenese und der Beziehung zur Entwicklung bei Pyrrhocoris apterus. *Zeitschr. f. wiss. Zool.* **51**, 685–736.
5. McClung CE (1899). A peculiar nuclear element in the male reproductive cells of insects. *Zool. Bull.* **2**, 187.
6. Paulmier F (1899). The spermatogenesis of *Anasa tristis. J. Morph.* **15** (Suppl), 223–272.
7. Sutton WS (1900). The spermatogonial divisions of *Brachystola magna. Kansas Univ. Quarterly* **9**, 73–100.
8. McClung CE (1902) The accessory chromosome – sex determinant? *Biol. Bull.* **3**, 43–84.
9. Stevens NM (1905). Studies in Spermatogenesis, with Especial Reference to the 'Accessory Chromosome'. Carnegie Institute, Washington, Pub. 36.
10. Wilson EB (1905). Studies on chromosomes. The behaviour of the idiochromosomes in *Hemiptera. J. Exp. Zool.* **2**, 371–405.
11. Wilson EB (1910). Studies on chromosomes. The chromosomes in relation to the determination of sex. *Science* **4**, 570–592.
12. Painter TS (1921). The Y chromosome in mammals. *Science* **53**, 503–504.
13. Painter TS (1923). Studies in mammalian spermatogenesis. II. The spermatogenesis of man. *J. Exper. Zool.* **37**, 291–335.
14. Barr ML and Bertram EG (1949). A morphological distinction between the neurones of the male and female, and the behaviour of the nucleolar satellite during accelerated nucleoprotein synthesis. *Nature* **163**, 676–677.
15. Moore KL, ed (1966). *The Sex Chromatin.* Philadelphia, WB Saunders.
16. Turner HH (1938). A syndrome of infantilism, congenital webbed neck and cubitus valgus. *Endocrinology* **23**, 566–574.
17. Polani PE, Hunter JF and Lennox B (1954). Chromosomal sex in Turner's syndrome with coarctation of the aorta. *Lancet* **2**, 120–121.
18. Polani PE, Lessof MH and Bishop PMF (1956) Colour blindness in ovarian agenesis (gonadal dysplasia). *Lancet* **271**, 118–120.
19. Ford CE, Jones KW, Polani PE, de Almeida JC and Briggs JH (1959). A sex chromosome anomaly in a case of gonadal dysgenesis (Turner's syndrome). *Lancet* **1**, 711–713.
20. Klinefelter HF, Reifenstein EC and Albright F (1942). Syndrome characterized by gynecomastia, aspermatogenesis without aleydigism, and increased excretion of follicle stimulating hormone. *J. Clin. Endocrinol. Metab.* **2**, 615–627.
21. Bradbury JT, Bunge RG and Boceabella RA (1956). Chromatin test in Klinefelter's syndrome. *J. Clin. Endocrin. Metab.* **16**, 689.
22. Ferguson-Smith MA and Munro IB (1958). Spermatogenesis in the presence of female nuclear sex. *Scot. Med. J.* **3**, 39–42 .
23. Ford CE, Jacobs PA and Lajtha LG (1958). Human somatic chromosomes. *Nature* **181**, 1565–1568.
24. Ford CE, Polani PE, Briggs JH and Bishop PMF (1959). A presumptive XXY/XX mosaic. *Nature* **183**, 1030–1032.
25. Jacobs PA, Baikie AG, Court Brown WM, MacGregor TN, MacLean N and Harnden DG (1959). Evidence for the existence of the human 'superfemale'. *Lancet* **2**, 423–425.
26. Jacobs PA (1982). The William Allan Award Memorial Address: Human population cytogenetics: the first twenty-five years. *Am. J. Hum. Genet.* **34**, 689–698.
27. Jacobs PA and Strong JA (1959). A case of human intersexuality having a possible XXY sex-determining mechanism. *Nature* **183**, 302–303.
28. Jacobs PA, Harnden DG, Court Brown WM, Goldstein J, Close HG MacGregor TN, MacLean N and Strong, JA (1960). Abnormalities involving the X chromosome in women. *Lancet* **1**, 1213–1216.
29. Ferguson-Smith MA and Johnston AW (1960). The human chromosomes in disorders of sex differentiation. *Trans. Assoc. Am. Phys.* **73**, 60–71.
30. Muldal S and Ockey CH (1960). The 'double male'. A new chromosome constitution in Klinefelter's syndrome. *Lancet* **2**, 492–493.
31. MacLean N, Mitchell JM, Harnden DG *et al.* (1962). A survey of sex chromosome abnormalities among 4514 mental defectives. *Lancet* **1**, 293.
32. Jacobs PA, Brunton M, Melville MM, Brittain RP and McClemont WF (1965). Aggressive behaviour, mental sub-normality and the XYY male. *Nature* **208**, 1351–1352.
33. Evans P (1982). Paul Polani. In: Adinolfi M, Benson P, Gianelli F, Seller M (eds). *Paediatric Research: a Genetic Approach.* London, Heinemann, pp. vii–x.

Addendum 1

Ford CE et al. (1959). A sex chromosome anomaly in a case of gonadal dysgenesis (Turner's Syndrome). *Lancet*, **1**, 711–713.
Reproduced with permission from Elsevier.

A SEX-CHROMOSOME ANOMALY IN A CASE OF GONADAL DYSGENESIS (TURNER'S SYNDROME)

C. E. FORD
Ph.D. Lond.

K. W. JONES
Ph.D. Wales

OF THE MEDICAL RESEARCH COUNCIL RADIOBIOLOGICAL RESEARCH UNIT, ATOMIC ENERGY RESEARCH ESTABLISHMENT, HARWELL, BERKS

P. E. POLANI
M.D. Pisa, M.R.C.P., D.C.H.

J. C. DE ALMEIDA
M.D. Brazil

J. H. BRIGGS
M.B. Lond., M.R.C.P.

OF GUY'S HOSPITAL, LONDON, S.E.1

GONADAL dysgenesis (ovarian agenesis, gonadal dysplasia) is a clinical syndrome usually presenting as a failure of secondary sex characteristics at puberty in girls whose gonads are absent or rudimentary. It is often associated with other congenital malformations such as small stature, digital anomalies, and, more rarely, webbed neck, congenital heart-disease, renal anomalies, intellectual subnormality, and other developmental errors. The more extreme expressions are often referred to as Turner's syndrome.

A considerable proportion of patients with gonadal dysgenesis are chromatin-negative (Décourt, Sasso, Chiorboli, and Fernandes 1954; Polani, Hunter, and Lennox 1954; Wilkins, Grumbach, and Van Wyck 1954), although chromatin-negativity (Barr and Bertram 1949) is an invariable feature of normal males. One possible explanation is that gonadal dysgenesis in man is due to castration while an embryo, since the experimental castration of embryonic rabbits results in the production of animals of female phenotype irrespective of the genetic sex-constitution of the embryo (Jost 1947). However, in abnormal individuals chromatin negativity or positivity ("nuclear sexing") may not necessarily indicate true chromosomal sex (*Lancet* 1956, Polani, Lessof, and Bishop 1956). An alternative explanation for the findings in gonadal dysgenesis might be abnormal sex differentiation following anomalous sex determination in the zygote.

Two approaches to the problem of certainly identifying the sex chromosomes present in Turner's syndrome suggested themselves: direct cytological observation, and the study of colour-blindness, which is a sex-linked recessive character and an X-chromosome marker. The results obtained by the second method agreed with the simple interpretation of the "nuclear sexing" results: chromatin-negative patients with gonadal dysgenesis seemed to have only one X chromosome (Polani et al. 1956). The presence or absence of the Y chromosome could not be determined and it was thought likely that the patients had an XY sex-chromosome constitution, although the possibility that they might be XO was also considered. Danon and Sachs (1957) also suggested that some patients with gonadal dysgenesis might have an XO sex-chromosome constitution, but that other patients might be examples of somatic mosaicism in respect of their sex-chromosome constitution. A study of the blood-groups of three patients with gonadal dysgenesis (Platt and Stratton 1956) supplied evidence that these individuals were not haploid—i.e., were not XO merely because all their chromosomes were unpaired.

Technical developments have recently made it possible to obtain accurate information regarding the somatic chromosomes of human patients, either in bone-marrow cells briefly incubated in vitro (Ford, Jacobs, and Lajtha 1958) or in cells from tissue cultures (Tjio and Puck 1958). In consequence the normal number of human chromosomes and their normal morphology are now reasonably well known. The subject of this report is a chromatin-negative case of Turner's syndrome whose bone-marrow cells proved to contain 45 chromosomes only, instead of the normal number of 46, and whose sex-chromosomal constitution is determined to be XO.

Case-report

The patient presented at the age of 14 with a short stature, primary amenorrhœa, and absence of secondary sex characteristics. In addition she was backward at school.

Family history.—Parents healthy. Father 5 ft. 4 in., mother 5 ft. 2 in. A maternal aunt, who died at the age of 21, was dwarfed and had had only two scanty periods; she was known to have pernicious anæmia. Two brothers 6 and 10 years old and one sister aged 9 were all healthy.

Personal history.—Maternal health good during pregnancy, and delivery normal. Birth weight 5 lb. 4 oz. Early development normal.

On examination, height 51 in., lower segment 25 in., arm-span 50 in. Weight 4 st. 13 lb.

There was slight facial asymmetry, low implantation of the ears, and a small chin. There was a high arched palate, a short broad neck without webbing, slight funnel deformity of the chest, cubitus valgus, pes cavus, and digital deformities. Cardiovascular system normal. Blood-pressure 110/70. Normal femoral pulses. There was no evidence of puberty.

Investigations

Examination of skin biopsy and blood smear showed a chromatin-negative pattern.

Follicle-stimulating hormone positive to 32 mouse units 17-ketosteroids 10·8 mg. per day.

Radiographically chest, heart, intravenous pyelography normal. Radiological assessment of bone age corresponded to the chronological age.

No defect in colour-vision (Ishihara) in patient or parents.

The marrow cells, obtained by a routine marrow puncture, were suspended in a mixture of glucose-saline and serum from the patient herself, and were sent to Harwell for cytological processing. After incubation the cells were exposed to colchicine for one hour, then fixed and stained by the Feulgen procedure. Squash preparations were made and the chromosomes were studied in cells arrested in the metaphase of mitosis by the action of the colchicine.

The chromosomes were counted in 102 cells of which 99 cells were found to have 45 chromosomes only. The remaining 3 cells contained fewer than 45 chromosomes and previous experience suggests that the deficiency is likely to be a consequence of damage to the cells during the making of the preparations. 14 cells were selected for detailed study. In every one of them 4 small acrocentric chromosomes were present, as in a normal female: in a normal male there are 5 of these chromosomes, one being the Y chromosome. All the selected cells also contained 15 medium-length metacentric chromosomes, as in a normal male—a normal female having 16 which include the two X chromosomes (Ford et al. 1958). These observations of themselves strongly suggest that the chromosome constitution is XO.

The individual recognition of the X and Y chromosomes may be a matter of some difficulty. However Tjio and Puck (1958) assert that X and Y chromosomes

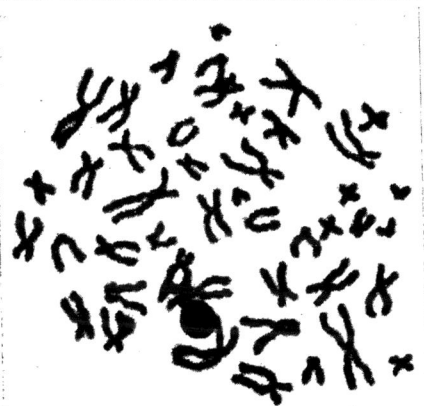

Fig. 1—Chromosomes (45) of the patient with Turner's syndrome discussed in the text. Colchicine-arrested metaphase in a bone-marrow cell. Feulgen squash preparation (× 2200). The round black body is probably an oil droplet.

can be recognised individually in their preparations made from tissue cultures. We agree that the Y chromosome can be distinguished in favourable cells of normal males, but we have not yet been able to identify the X chromosome (or chromosomes) unequivocally in bone-marrow preparations. Nevertheless in many of the selected cells of the present patient it was possible to make a reasonably satisfactory classification of the chromosomes into 22 pairs and one odd chromosome. A photograph of one of these cells is reproduced in fig. 1. In fig. 2 the chromosomes from the same cell are shown arranged in pairs. Suspicion that the odd chromosome is the X is inevitable, but this chromosome and the two members of pair 6 are very similar in length and arm-ratio and their true relationships remain uncertain. The probability that one of the three is X is strengthened by the good agreement of their proportions with those of undoubted X-chromosomes in X-Y bivalents at metaphase in primary spermatocytes (Ford and Hamerton 1956). Experience of numerical chromosomal abnormalities in animals and plants (Swanson 1957) suggests that it is very improbable that a human individual who has only 45 chromosomes as the result of the loss of a large or medium-sized autosome would be viable, but that an XO zygote might well develop to maturity. We therefore conclude that the sex-chromosome constitution of the patient is XO.

Discussion

These observations are of interest not only with reference to Turner's syndrome. Here is an individual who is female anatomically and psychologically, whose cells are "male" as judged by nuclear sexing, and whose chromosomes are neither normally male nor normally female. However, as judged by her chromosomes she has no male component, but half a normal female component, and there seems no justification for considering her to be really male in any sense. It must therefore be accepted that chromatin negativity does not necessarily imply maleness and it would probably be best if the phrase "nuclear sexing" were dropped from the vocabulary and the more accurate if less striking terms, chromatin negativity or positivity, were always used instead. The very real clinical reasons for doing this have already been stressed.

An explanation of the origin of the sex chromosome anomaly in gonadal dysgenesis can be sought in the process of non-disjunction, best known as an abnormality of oogenesis in *Drosophila melanogaster* (Morgan, Bridges, and Sturtevant 1925). Non-disjunction of the sex-chromosomes in the female fly implies the migration of two X chromosomes to one pole of the spindle during one of the two meiotic anaphases. Thus the ovum comes to contain either the haploid number of autosomes plus *two* X chromosomes, or only the haploid number of autosomes *without* X chromosomes. Fertilisation of an ovum of the latter type by a Y-bearing sperm results in a non-viable YO zygote; fertilisation by an X-bearing sperm yields an XO zygote which develops into a sterile female. Our findings suggest that in man an XO zygote develops into a sterile "agonadal" individual whose phenotype is female.

Fertilisation of the other type of abnormal ovum (XX) by an X-bearing sperm gives, in drosophila, an XXX zygote with poor viability, but which occasionally survives pupation and then emerges as a fly with accentuated female secondary sexual characteristics, technically called a "super-female"; fertilisation by a Y-bearing sperm yields an XXY zygote which develops into a fertile female. In man the XXX state is as yet unknown, but evidence that the XXY individual appears as a chromatin-positive case of Klinefelter's syndrome has been presented (Ford, Polani, Briggs, and Bishop 1959, Jacobs and Strong 1959.) The evidence in favour of

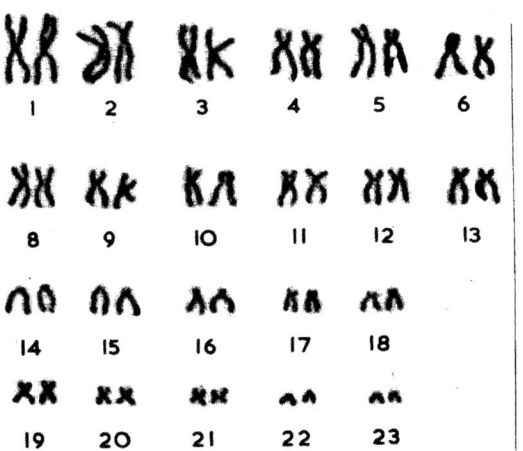

Fig. 2—Chromosomes from the cell shown in fig. 1 arranged in pairs (× 2200).

the occurrence of non-disjunction in man is thereby strengthened.

Non-disjunction has so far been considered as occurring during oogenesis only. Family colour-vision studies in cases of gonadal dysgenesis suggest that it may also occur during spermatogenesis. If an O ovum is fertilised by an X sperm and the resultant individual has a major red-green colour-vision defect, the father should also show the colour-vision defect. But in four families where patients with gonadal dysgenesis (chromatin-negative) have a major red-green colour-vision defect this was not the case (Lenz 1957, Stewart 1958 personal communication; and two of our families, see Bishop et al. 1959). It would appear that in these patients the X-chromosome with the anomalous colour-vision gene was not of paternal origin but was derived from a heterozygous (carrier) mother. These patients have not been examined cytologically, but, if they are XO, it will follow that they developed from zygotes arising from the fertilisation of normal X-bearing ova by sperm carrying neither X nor Y. Such sperm would arise as a result of non-disjunction during spermatogenesis. Evidence that this is by no means unlikely is provided by the observation that X and Y chromosomes are sometimes unpaired at metaphase in first spermatocytes (Ford and Hamerton 1956). Both chromosomes would then be expected to migrate to the same pole in approximately 50% of the ensuing anaphases.

In conclusion it should be emphasised that the XO patient should not be referred to as an instance of "sex-reversal", as a "chromosomal male", or as a "genetic male": she is a female, with an abnormal genotype.

We wish to thank Dr. P. M. F. Bishop for permission to study a patient under his care. We acknowledge the skilful technical assistance of Mr. G. D. Breckon, Miss P. A. Moore, and Miss S. R. Wakefield.

REFERENCES

Barr, M. L., Bertram, E. G. (1949) *Nature, Lond.* **163**, 676.
Bishop, P. M. F., Lessof, M. H., Polani, P. E. (1959) Memoir no. 7. Society for Endocrinology, London. Ed.: C. R. Austin (in the press).
Danon, M., Sachs, L. (1957) *Lancet*, ii, 20.
Décourt, L., Sasso, W. da S., Chiorboli, E., Fernandes, J. M. (1954) *Rev. Assoc. med. Brazil*, **1**, 203.
Ford, C. E., Hamerton, J. L. (1956) *Nature, Lond.* **178**, 1020.
— Jacobs, P. A., Lajtha, L. G. (1958) *ibid.* **181**, 1565.
— Polani, P. E., Briggs, J. H., Bishop, P. M. F. (1959) *ibid.* (in the press).
Jacobs, P. A., Strong, J. A. (1959) *ibid.* **183**, 302.
Jost, A. (1947) *C.R. Soc. Biol., Paris*, **141**, 126.
Lancet (1956) **2**, 127.
Lenz, W. (1957) *Acta Geneticæ Medicæ et Gemellologiæ*, **6**, 231.
Morgan, T. H., Bridges, C. B., Sturtevant, A. H. (1925) *Bibliog. Genetica*, **2**, 1.
Platt, R., Stratton, F. (1956) *Lancet*, ii, 120.
Polani, P. E., Hunter, W. F., Lennox, B. (1954) *ibid.* ii, 120.
— Lessof, M. H., Bishop, P. M. F. (1956) *ibid.* ii, 118.
Stewart, J. S. S. (1958) Personal communication.
Swanson, C. P. (1957) Cytology and Cytogenetics. (Prentice Hall, Englewood Cliffs, N.J.).
Tjio, J. H., Puck, T. T. (1958) *Proc. Nat. Acad. Sci*, **12**, 1229.
Wilkins, L., Grumbach, M. M., Van Wyck, J. J. (1954) *J. clin. Endocrin.* **14**, 1270.

Addendum 2

Jacobs PA and Strong JA (1959). A case of human intersexuality having a possible XXY sex-determining mechanism. *Nature*, 183, 302–303.
Reproduced with permission from Nature Publishing Group.

A CASE OF HUMAN INTERSEXUALITY HAVING A POSSIBLE XXY SEX-DETERMINING MECHANISM

By PATRICIA A. JACOBS and Dr. J. A. STRONG

Medical Research Council Group for Research on the General Effects of Radiation and Department for Endocrine and Metabolic Diseases, Western General Hospital and University of Edinburgh

RECENT improvements in techniques for the examination of human somatic chromosomes have made possible the study of the chromosome complement of human intersexes; consequently, it is now practicable to investigate the relationship in these cases between sex as determined by direct chromosome study, and sex as inferred from the study of 'nuclear sex chromatin' of the type described by Barr and Bertram[1]. This report is concerned with one of a series of patients with gonadal dysgenesis who are under investigation, and the particular feature of interest is the occurrence of 47 somatic chromosomes in contrast with the normal number of 46 in man.

In recent years the diploid chromosome number of 46 has been recorded in a large number of instances. In addition to the 60 cases cited in a previous publication[2] we have recorded a diploid number of 46 in bone marrow preparations from a further 40 European subjects. Kodani, on the other hand, has recently published results of counts made on testicular material from 36 Japanese and 8 American white males[3,4]. He claims that in 16 Japanese and 1 American there was a diploid number of 48; in 2 Japanese a diploid number of 47, and that in the remaining 13 Japanese and 7 whites the number was 46. He suggests that 46 is the basic diploid chromosome number for man, but in some instances there are additional "supernumerary chromosomes". The occurrence of this type of supernumerary chromosome, however, has not been reported previously among the vertebrates and awaits confirmation by other workers.

Our patient, an apparent male aged twenty-four, was presented as a case of gonadal dysgenesis with gynæcomastia and small testes associated with poor facial hair-growth and a high-pitched voice. Biopsy examination of testicular tissue showed the seminiferous tubules to be extremely hyalinized and atrophic, and also an apparent increase in the number of interstitial cells. Chromosome studies were attempted on part of this material, but no spermatogonial mitotic or meiotic divisions were seen. Smears made from both the buccal mucosa and the blood were examined by Dr. B. Lennox of the Department of Pathology, Western Infirmary, Glasgow, and found to demonstrate typical female morphology with regard to their nuclear sex chromatin.

Material obtained by sternal marrow puncture was used for investigating the somatic chromosomes. The technique used for culturing the material in the presence of colchicine and for making squash preparations has already been described[2].

The chromosomes were counted in 44 cells in metaphase and the results are shown in Table 1.

The majority of the cells contained 47 chromosomes, and in all those cells where the chromosomes were well fixed and spread, the count was undoubtedly 47

Table 1

Chromosome No.	45	46	47	48	49
No. of cells	2	7	29	5	1

(Fig. 1). The apparent variation is in all probability due to technical errors. Fragments of cells containing chromosomes may become lost during the squashing process so that counts lower than the diploid number are obtained; and occasionally chromosomes split at the centromere and individual chromatids may be counted as chromosomes, giving a count higher than the diploid number[2].

A study of the chromosome morphology in 8 cells of a suitably high standard showed that each of these had a normal male complement with the Y chromosome present, but that there was also an extra chromosome having a sub-median centromere occurring in the medium size range. In the normal male there are 15 chromosomes in this range, and in the female 16, all having sub-terminal or sub-median centromeres[2]. Owing to the slight variations in their size and morphology these chromosomes have proved difficult to pair, and it is in this category that the X chromosome is to be found.

There are strong grounds, both observational and genetic[5,6], for believing that human beings with chromatin-positive nuclei are genetic females having two X chromosomes. The fact that this patient is chromatin-positive and has an additional chromosome within the same size range as the X, as well as an apparently normal Y, makes it seem likely that he has the genetic constitution XXY. The possibility cannot be excluded, however, that the additional chromosome is an autosome carrying feminizing genes.

The presence of the extra chromosome might have been due to one or other of the parents having 47 chromosomes, and, therefore, chromosome studies were made on marrow specimens from both parents. Both were found to have a diploid number of 46 (Table 2), and analysis of cells of suitable quality showed the morphology of the chromosomes to be normal.

The occurrence of the extra chromosome therefore may be due to non-disjunction at either mitosis or meiosis during gametogenesis in one or other parent. Alternatively, it may be due to non-disjunction

Fig. 1. Metaphase plate showing 47 chromosomes

Table 2

Father	Chromosome No.	44	45	46	47	48	49
32 cells counted	No. of cells	2	3	26	—	—	1
Mother	Chromosome No.	44	45	46	47	48	49
39 cells counted	No. of cells	1	3	33	2	—	—

occurring during the patient's very early embryological development, in which case there is a possibility that the patient may be a mosaic. Unfortunately, it is not possible with the techniques at present available to examine the chromosomes of tissues arising from different germ layers of the embryo.

We would like to thank Dr. B. Lennox of the Department of Pathology, Western Infirmary, Glasgow, for checking the nuclear sex of the preparations of buccal mucosa and blood and Miss M. Brunton for technical assistance.

A further report of this and other cases will follow.

[1] Barr, M. L., and Bertram, E. G., *Nature*, **163**, 676 (1949).
[2] Ford, C. E., Jacobs, P. A., and Lajtha, L. G., *Nature*, **181**, 1565 (1958).
[3] Kodani, M., *Proc. U.S. Nat. Acad. Sci.*, **43**, 285 (1957).
[4] Kodani, M., *Amer. J. Human Genetics*, **10**, 125 (1958).
[5] Grumbach, M. M., and Barr, M. L., "Recent Progress in Hormone Research", **14**, 255 (1958).
[6] Polani, P. E., Bishop, P. M. F., Lennox, B., Ferguson-Smith, M. A., Stewart, J. S. S., and Prader, A., *Nature*, **182**, 1092 (1958).

CHAPTER 5

The other autosomal trisomies, 1960

We have seen that the publication early in 1959 of the chromosomal basis of Down's syndrome by Lejeune et al.,[1-3] confirmed rapidly by others,[4,5] showed clearly that a human congenital malformation syndrome could be caused by a specific and consistent chromosome anomaly and that this brought human cytogenetics, for the first time, firmly into the orbit of diagnostic medicine. Now the question became: how many other such disorders might have underlying chromosomal defects or imbalances? Since few of those who developed the cytogenetic techniques for human chromosomes had themselves the necessary clinical, especially paediatric skills, new links had rapidly to be forged, often between workers who previously had little understanding of each other's disciplines. These collaborations varied in their closeness and duration, but were especially essential during these early years, when human cytogenetics had yet to become an established clinical discipline.

The search for new human chromosome disorders in fact took longer to achieve success than those involved expected – hardly surprising with hindsight, given that most laboratories remained dependent on autopsy or bone marrow samples, were able only to process small numbers, and could only confidently identify changes in chromosome number rather than in morphology. Even workers such as David Harnden, whose less invasive skin fibroblast culture techniques gave access to a wide range of patients, found a disappointingly high proportion of negative results, as seen below.

1960 was, though, marked by the discovery of two important new chromosomal syndromes, the autosomal trisomies now known as trisomies 18 and 13, though the precise chromosome involved was not certain initially. The nomenclature was also variable, and confusing to present day readers of the original papers; the eponymous terms 'Edwards' and 'Patau' syndromes, are less used today, but the workers involved, not just those reflected in the names, deserve to be remembered since they were pioneers both in human genetics and in the more clinical study of malformation syndromes

TABLE 5-1

KEY EARLY PAPERS ON HUMAN CYTOGENETICS PUBLISHED BY *THE LANCET*

Year	Authors	Chapter/reference	Topic
1954	Polani et al.	Ch 4[17]	Sex chromatin in Turner syndrome
1959	Jacobs et al.	Ch 3[15]	Trisomy 21
	Ford et al.	Ch 4[19]	Chromosomes in Turner syndrome
	Jacobs et al.	Ch 4[25]	XXX female
	Baikie et al.	Ch 6[10]	Chromosomes in leukaemia
1960	Polani et al.	Ch 3[22]	Translocation Down's syndrome
	Penrose et al.	Ch 3[23]	Familial Down's translocation
	Carter et al.	Ch 3[24]	Familial Down's translocation
	Edwards et al.	Ch 5[6]	Trisomy E (18)
	Patau et al.	Ch 5[7]	Trisomy D (13)
	Böök and Santesson	Ch 5[12]	Triploidy
	Jacobs et al.	Ch 4[28]	X chromosome abnormalities
	Denver conference	Ch 7[19]	Denver conference report

The two papers[6,7] appeared side-by-side in *The Lancet* on April 9th 1960 (this was unknown to each group until the journal actually appeared). The role of *The Lancet* in publishing a remarkably high proportion of discoveries in human cytogenetics at this time deserves a note here, since its wide international circulation among clinicians of all specialities must have considerably hastened the awareness of the medical community regarding the importance (or even existence) of chromosomes in relation to human disease. Several examples were encountered in the interview series where a senior clinician had directly suggested to a young investigator that chromosome studies should be started on the basis of one of the first *Lancet* papers. Until 1959 papers reporting discoveries in human cytogenetics had appeared mostly in relatively low-profile specialist journals, unless they made a particularly fundamental scientific point, when a journal such as *Nature* might accept them. In 1959, however, with clinical applications becoming clear, *The Lancet*, with its editorial office based in London, began to publish a series of key chromosome reports on human chromosomes; it had innovative and well informed editors and a rapid publication time, though of necessity papers had to be brief. Table 5-1 lists some of the most significant papers.

The influence of editorial and reviewing policies of journals on how these early developments in human genetics were disseminated and interpreted to the wider scientific and medical community has been considerable, and deserves study in its own right by science historians. *The Lancet* seems to have had a general policy of trying to encourage, identify and anticipate contributions in new but medically relevant fields, later diverting them to more specialist journals once the topic had become more familiar. It does not seem that the journal had a deliberate policy of favouring genetics *per se*, but rather that it recognised human cytogenetics at this time as a new growing point in medicine (I am grateful to David Sharp, formerly Deputy Editor of *The Lancet*, for discussion on this point). From the historical viewpoint the concentration of these key papers in journals such as *The Lancet* and *Nature* is helpful since most libraries retain complete series from the earliest years and back numbers are now being digitised.

The first of the two *Lancet* papers was on an infant girl from Birmingham, England, with a wide range of anomalies involving multiple systems, which proved rapidly fatal.[6] John Edwards, first author on the paper and both clinician and geneticist, had recently moved from Birmingham to Oxford, where he had close links with Charles Ford and his colleagues at nearby Harwell, who undertook the chromosome studies. Edwards, in discussion and interview in 2004,

JOHN H EDWARDS (born 1928)

Born in London, John Edwards trained in zoology and medicine at Cambridge University and King's College Medical School, London. He had a strong early interest in natural history and also in gliding and polar exploration, this last leading to work as a ship's surgeon in Antarctica soon after qualifying as a doctor. After developing tuberculosis he was advised to make a career in psychiatry rather than clinical medicine, but then took a post in epidemiology and genetics in Birmingham working under MacKeown and Lancelot Hogben on the epidemiology of neural tube defects.

In 1957 he moved to the newly established Medical Research Council Population Genetics Unit at Oxford, establishing close links with the Harwell cytogenetics group (Charles Ford and David Harnden), which he maintained after returning to a lectureship in Birmingham. It was at this point that he made his major contributions in identifying trisomy 18 and also X-linked hydrocephalus,[15] both published in 1960.

As Professor first in Birmingham and later in Oxford, he became particularly involved in the genetics of the HLA system and also in analytical approaches to human and comparative gene mapping, an interest continued to the present. (Photograph courtesy of Professor John Edwards.)

DAVID G HARNDEN (born 1932)

David Harnden was born in London but grew up in Edinburgh from an early age. He studied zoology at Edinburgh University, followed by a PhD in cell biology. In 1957 he went to work with Charles Ford at Harwell, first learning tissue culture techniques with Honor Fell at Cambridge. He carried out studies on many different congenital malformations and was responsible for the trisomy 18 discovery with John Edwards and the use of cultured skin fibroblasts for chromosome analysis.[16]

Returning to Edinburgh in 1959 he became part of the human cytogenetics group with Patricia Jacobs, whom he had already met at Harwell, and Michael Court Brown, being especially involved in the initial sex chromosome and leukaemia chromosome discoveries. Developing an interest in viral chromosome damage in relation to cancer following collaborative work with Howard Temin in America, he moved in 1969 to the Chair of Cancer Studies in Birmingham (UK), working in particular on ataxia telangiectasia and related chromosomal breakage syndromes. From 1983 until his retirement in 1997 he was head of the Paterson Institute for Cancer Research, Manchester. (Photograph courtesy of Professor David Harnden.)

played down his own role as 'being responsible for ensuring samples were placed in the right bottles and delivered to the right people' (in itself a far from unimportant task since in the event the duplicate set of samples processed in Edwards' own Oxford laboratory failed to grow in culture), but David Harnden, who undertook the cytogenetic study, also interviewed in 2004, was quite clear that John Edwards' clinical acumen had alerted him to the likelihood of a chromosome abnormality.

DH. *At that time John Edwards was sending in material from children and this little girl, he said, 'Now this really is a strange little girl. I think you are going to find something here', and when I looked down the microscope I found 47 chromosomes. It was really quite astonishing. I wasn't very sure whether it was chromosome 17 or chromosome 18 to be honest, and that was written up and published and it was back-to-back in* The Lancet *with a paper by Klaus Patau on Trisomy 13. It was quite astonishing....*

John was particularly excited about that because he said, 'You know if you think of the galaxy of abnormalities in Down's syndrome, this is a similar but different galaxy of abnormalities'. I'd looked at things like anencephaly and so on. I actually published a paper of all the negatives and that's in the archives somewhere, but this one was indicated to be special because John thought from a clinical point of view there were similarities with Down's syndrome.

PSH. *What was your reaction when you found there was an extra chromosome?*

DH. *I guess I was astonished because I'd looked at quite a lot of negatives up until then. Not only anencephaly, but hydatidiform moles and all sorts of things. But that was very, very exciting.*

John Edwards likewise confirmed that it was the particular combination of clinical features that made him suspect a chromosome abnormality.

JHE. *Well, I had been reading about these things and so on and I was particularly interested in the Datura plant which has 12 chromosomes and 12 trisomic syndromes, all of which have their features that are disproportionate. Everything is disproportionate but nothing is very critical because I suppose it wouldn't be alive.....*

PSH. *So you came across this child, was it at one of the meetings that this came up for discussion?*

JHE. *Yes, I was very fortunate actually ... there was this case, which I was told, 'Oh an interesting case that we have got, I am sure you would like to see this case of Ullrich'. ...*

There was this strange looking child and, as with Down's syndrome, everything was wrong, but nothing very wrong. Well it was worse than Down's syndrome and so I thought, I did actually think, this is what a trisomy ought to be like, so I could claim to have made a diagnosis of 'trisomy of an unknown nature.' I was just developing this skin biopsy procedure and didn't like to do it until I had had a bit more experience; it is very simple taking these small skin biopsies, but I didn't like to do it on this dying child. So I said, as soon as the child is dead I will rush up to the autopsy. I took some hand prints and looked at the child in great detail, and then I went back; the next thing I knew it had died. Friday afternoon was the autopsy and it was foggy and miserable and I was driving up to this autopsy, which of course was done beautifully by Hugh Cameron. I took double samples of everything and then I drove back to Oxford and gave half of it to the person who was doing the chromosomes there, but in fact I knew that probably it wouldn't do very well because there was a very high infection rate in cultures; the other half I rang up Charles Ford on Saturday morning and tried to talk him into doing this and he said, well he would go and see David Harnden and he agreed to come out. So there were all these little tubes I'd got with all sorts of things marked, with tissues, and David set them all up and the winner was lung. I have never heard of anyone growing lung, but anyway in 10 days David had the most beautiful preparations and the Oxford ones all died. It was actually diagnosed as trisomy 17 by our experts; they were beautiful chromosomes of course. I had hardly ever seen any before, but in their view it was 17. Charles, with his usual generosity, didn't put his name on the paper because he thought David should get full credit. Hugh Cameron certainly deserved it because he made some very sound observations which were based on very large experience. Anyway, this eventually got published, as trisomy 17 I think.

Edwards is right to pay tribute to the contribution of Hugh Cameron, paediatric pathologist in Birmingham, in ensuring that the different malformations were expertly documented at autopsy. This, like the detailed clinical observations, forms a vital part of such research studies and is often under-recognised.

The published paper indeed refers to the trisomy as 17, not 18 (see Figure 5-1 and

Fig. 5-1 Chromosomes of trisomy 18 patient from Edwards *et al.* (paper reproduced at the end of this chapter, courtesy of *The Lancet*). Note that the extra chromosome is identified here as 17 rather than 18 (see text).

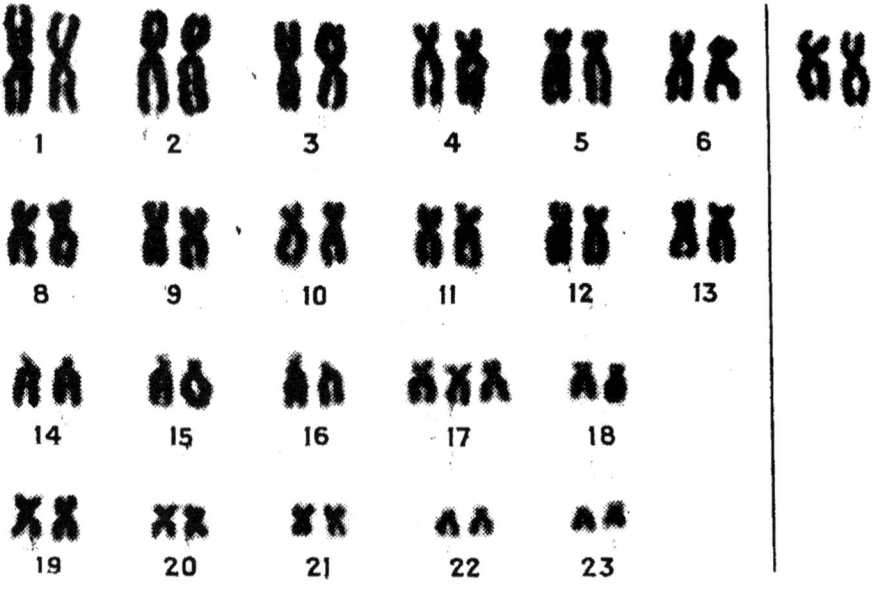

the Edwards *et al.* 1960 paper reproduced at the end of this chapter); until nomenclature had been standardised by the Denver conference later in 1960 (see Chapter 7), each group had its own criteria for identifying specific chromosomes.

The clinical criteria for suspecting an autosomal trisomy, described above by both Edwards and Harnden, are well summarised in the opening paragraph of *The Lancet* paper. The likelihood of lethality is also noted.

There are 22 pairs of autosomes in man, and it is possible that 22 syndromes of autosomal trisomy exist; some, however, may be incompatible with intra-uterine life. Mongolism is due to trisomy of one of the smallest chromosomes (Lejeune et al. 1959) and, as it seems reasonable to suppose that the larger the extra autosome the worse the disturbance, most other trisomic syndromes might be expected to be at least as severe. In man such aberrations would probably lead to generalised disorders of shape and size and to multiple structural abnormalities rather than to a single localised deformity.

The second of the pair of *Lancet* reports, on what is now known as trisomy 13, came from the University of Wisconsin, Madison, and again was on a multiply malformed child (Figure 5-2). Here, the cytogenetics workers were Drs Klaus Patau and Eeva Therman, both post-World War II

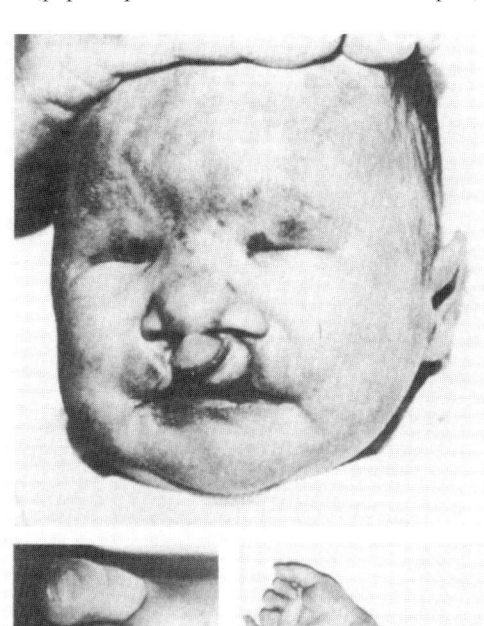

Fig. 5-2 Clinical photograph of original trisomy 13 patient. From Therman, 1980;[18] reproduced with permission from Springer Science and Business Media. This photograph also appears in Patau *et al.* (paper reproduced at the end of this chapter).

immigrants to America from Europe, and pioneers of human cytogenetics in the United States. The clinician involved, David Smith, later became the founder of the discipline of clinical dysmorphology. Sadly, none of the authors are now living (Eeva Therman died in 2004, shortly before an interview could be arranged), but brief biographical information and further sources are given in the boxes. The eponym

KLAUS PATAU (1908–1975)

Born in Gelsenkirchen, Germany, Klaus Patau received his PhD from the University of Berlin in 1936, the subject being comparison of banding patterns in the giant salivary gland chromosomes in different species of *Drosophila*, work which gave him an international reputation. He spent a year, 1938–9, at the John Innes Horticultural Institute in the UK, but at the outbreak of war was drafted into the German army and wounded in the Russian campaign, before returning to work in Berlin with the Russian geneticist Timofeeff-Resovsky.

After the war Patau left Germany, coming first to the UK and then to the United States, where he was given a post by the University of Wisconsin, Madison, remaining there for the rest of his life. In 1959 he joined the new medical genetics department formed by Dr James Crow, making an immediate impact on human cytogenetics with his work on the human autosomal trisomies, and the development of diagnostic cytogenetics, including its nomenclature.

He is today remembered for his work on human chromosomes, but it could be argued that his earlier work on heterochromatin, the sex chromosomes, and on measurement of DNA in chromosomes,[17] is of equal scientific importance. A short biographical memoir has been written by colleagues at University of Wisconsin. (Unpublished note from University of Wisconsin records kindly provided by Dr James F Crow; photograph kindly provided by Dr Helena Pihko, Helsinki.)

'Patau' syndrome was suggested after Patau's death and according to Dr John Opitz, (in Madison as a colleague from shortly afterwards), this gave rise to some friction as to recognition, so it is perhaps fortunate that the simple and more specific title 'trisomy 13' is now generally used.

Dr James Crow, colleague of Klaus Patau and Eeva Therman, and founder of the Madison Medical Genetics Department, has given the background to the study that led to discovery of trisomy 13 (and which also independently detected a patient with trisomy 18).[8]

At that time Klaus and I regularly had lunch together, usually with Eeva. One day I attended a seminar in the Paediatrics Department at which I heard of Lejeune's discovery of trisomy 21. I immediately told Klaus and he agreed to look for other trisomics.

We reasoned, partly by analogy with trisomics in Datura, that any trisomic

EEVA THERMAN (1916–2004)

Born in Helsinki and trained originally in botany, Eeva Therman came to study cytogenetics in the University of Wisconsin, Madison, in 1958, joining Klaus Patau, and was directly responsible for the observations on the human trisomies published in 1960 and thereafter. She married Patau in 1961 and continued her cytogenetics research after his death in 1975, making particular contributions to the structure of the X chromosome.

Today Eeva Therman is probably best known for her widely read book *Human Chromosomes*,[18] later co-authored by O.J. Miller and still in print. Towards the end of her life she returned to her native Helsinki, where she died in 2004. (Photograph kindly provided by Dr Helena Pihko, Helsinki.)

would have several abnormalities and very likely mental retardation would be included. So Klaus decided to look in the Wisconsin Central Colony for children with, in addition to mental retardation, two unrelated abnormalities. He enlisted the help of David Smith to help with the clinical aspects.

Surprisingly, they found one on the very first day. This was a child with cleft lip and polydactyly and turned out to be trisomy 13. The next day they found trisomy 18. I remember our thinking during lunch that, at this rate, he would have all 22 autosomal trisomies in less than a month. But, as you know these two were all (except for sex chromosomes and the unusual things found later).

The source of patients in this study explains why both the trisomy 13 and 18 patients were living at an older age than usual, since only survivors, rather than sick newborns, would have been at such an institution. According to John Opitz (personal communication 2005) the trisomy patients continued to live, though severely mentally and physically handicapped, for a number of years afterwards.

The paper of Patau et al. emphasises the likelihood of multiple abnormalities occurring in any autosomal trisomy and the possibility that many might not be viable.[7] It also makes the far-sighted prediction that chromosome abnormalities would become important in human gene mapping on account of the specific abnormalities indicating the involvement of relevant genes in that chromosomal region.

It was to be expected that other autosomal trisomics, if they should be at all viable, would also display multiple congenital disturbances.

DAVID W SMITH (1926–1981)

David Smith was born in Oakland, California, and trained in paediatrics. In 1958 he joined the University of Wisconsin Medical School, Madison, and in 1960 was responsible for the clinical delineation of the first autosomal trisomy patients as part of the cytogenetic study led by Klaus Patau.

In 1966 he became Head of Paediatrics at University of Washington, Seattle, where he systematically developed studies of human malformations and became the founder of the field known today as clinical dysmorphology. While not trained in genetics, David Smith had a pivotal influence through his training of numerous clinicians who would continue the development of research in this area and through his widely used book, continued up to the present time by colleagues, *Recognisable Patterns of Human Malformation*.[19] Following his early death, the annual 'David Smith workshops' both preserve his name and continue to foster this field of work. (Photograph kindly provided by Dr Judith Hall, Vancouver.)

Of the 22 conceivably viable types of autosomal trisomics three have been found so far. How many more are viable is not known, but it seems certain that not all of them are. As each of the viable ones will probably show a clinically distinguishable syndrome, we may expect that an aetiologically unique group of a limited number of 'autosomal trisomy syndromes' will become established. To the human geneticist these will be of continuing interest; he will in particular look forward to cases in which the presence of known genes in the parents can be related to peculiarities in the trisomic child. It seems likely that every component anomaly of a trisomy syndrome reflects, by way of a dose effect of gene action, the presence in the respective chromosomes of at least one gene locus that plays a prominent role in the normal development of the afflicted organ; indeed, we suspect that many of the trisomy anomalies can also be produced individually in euploid persons by the heterozygous or homozygous presence of a suitable mutated allele of the responsible gene. Polydactyly and hare lip with cleft palate in the present patient may be cases in point. These also suggest that trisomy may become instrumental in establishing for the first time autosomal linkage groups in man

The trisomy 13 discovery resulted in a detailed, clinically orientated publication, in addition to the shorter *Lancet* paper, with

Fig. 5-3 The 12 phenotypes corresponding to the 12 possible trisomies in the thorn apple, *Datura stramonium*; reproduced from Hurst, 1932, after Blakeslee, 1929. Blakeslee's observations on the *Datura* trisomies had a profound influence on the later search for human autosomal trisomies.

David Smith as first author.[9] Quite apart from this solution allowing credit to be spread more widely, this policy is also important in scientific terms, since it allows the clinical basis of the disorder to be documented in detail, rather than relegated to a few paragraphs (or even sentences) as is all too often the case today for high profile and especially basic scientific journals.

Of equal importance to the clinical insights was the awareness of both groups of the importance of the model for trisomies provided by studies of plant chromosomes, in this case the thorn apple, *Datura*, the work referred to probably being that of Blakeslee's 1934 paper,[10] and a series of previous papers from 1927. Blakeslee's work had a considerable influence on geneticists from the late 1920s onwards and a plate showing the different trisomic phenotypes for each of the 12 chromosome pairs is shown in Figure 5-3. Few human geneticists today are aware that the effects of trisomy were so clearly laid out 80 years ago.

By the end of 1960 it was becoming clear that congenital malformations or other constitutional disorders involving gain (or loss) of a complete autosome were either absent or at least excessively rare in liveborn children, and that for the great majority of malformations one would have either to search for more detailed alterations using improved techniques, or indeed to look outside chromosomal causes. It also seemed likely that the finding that all the three recognised autosomal trisomies involved some of the smallest chromosomes, 21, 18 and 13, and that the last two disorders were in most cases rapidly fatal, was indicating that most other trisomies and corresponding aneuploidies were likely to be lethal before birth. A hint of this was soon suggested by the finding of triploidy in a spontaneous abortion by Penrose and Delhanty;[11] a triploid patient had already been described by Böök and Santesson in 1960,[12] but there are puzzling aspects to this case, whose clinical features were relatively mild and who may have been a mosaic, since blood cultures studied some years later showed a normal karyotype.[13] It would be several more years until the systematic study of Carr in 1965[14] confirmed both the range and remarkably high frequency of chromosome abnormalities in spontaneous abortions (see Chapter 8), thus bringing cytogenetics into the clinical practice of the gynaecologist, as well as of the paediatrician and endocrinologist.

References

1. Lejeune J, Gautier M and Turpin R (1959). Les chromosomes humains en culture de tissus. *C. R. Acad. Sci.* **248**, 602–603.
2. Lejeune J, Gautier M and Turpin R (1959). Étude des chromosomes somatiques de neuf enfants mongoliens. *C. R. Acad. Sci.* **248**, 1721–1722.
3. Lejeune J, Turpin R and Gautier M (1959). Le mongolisme, maladie chromosomique. *Bull. Acad. Nat. Méd.* **143**, 256–265.
4. Jacobs PA, Baikie AG, Court Brown WM and Strong JA (1959). The somatic chromosomes in mongolism. *Lancet* **1**, 710.
5. Böök JA, Fraccaro M and Lindsten J (1959). Cytogenetical observations in mongolism. *Acta Paediatr.* **48**, 453–468.
6. Edwards JH, Harnden DG, Cameron AH, Crosse VM and Wolff OH (1960). A new trisomic syndrome. *Lancet* **1**, 787–790.
7. Patau K, Smith DW, Therman E, Inhorn SL and Wagner HP (1960). Multiple congenital anomaly caused by an extra autosome. *Lancet* **1**, 790–793.

8. Patau K, Therman E, Smith DW and Demars RI (1961). Trisomy for chromosome No. 18 in man. *Chromosoma* **12**, 280–285.
9. Smith DW, Patau K, Therman E and Inhorn SL (1960). A new autosomal trisomy syndrome. Multiple congenital anomalies caused by an extra chromosome. *J. Pediatr.* **57**, 338–345.
10. Blakeslee AF (1934). New Jimson weeds from old chromosomes. *J. Hered.* **25**, 81.
11. Penrose LS and Delhanty JDA (1961). Triploid cell cultures from a macerated foetus. *Lancet* **1**, 1261–1262.
12. Böök JA and Santesson B (1960). Malformation syndrome in man associated with triploidy (69 chromosomes). *Lancet* **1**, 858–859.
13. Böök JA, Masterson JG and Santesson B (1962). Malformation syndrome associated with triploidy – further chromosome studies of the same patient and his family. *Acta Genet.* **12**, 193–201.
14. Carr D (1965). Chromosome studies in spontaneous abortions. *Obstet. Gynaecol.* **26**, 306–326.
15. Edwards JH, Norman RM and Roberts JM (1960). Sex-linked hydrocephalus. Report of a family with 15 affected members. *Arch. Dis. Child* **36**, 481–485.
16. Harnden DG (1960). A human skin culture technique used for cytological examination. *Br. J. Exper. Pathol.* **41**, 31–37.
17. Patau K and Swift H (1953). The DNA content (Feulgen) of nuclei during mitosis in a root tip of onion. *Chromosoma* **6**, 149–169.
18. Therman E (1980). *Human Chromosomes*. New York, Springer.
19. Smith DW (1974). *Recognizable Patterns of Human Malformation*. Philadelphia, WB Saunders (2005 edition edited by K Jones).

Addendum

The two original papers on the autosomal trisomies, reproduced from *The Lancet* (1960) 1, 787–790 and 790–793.
Reproduced with permission from Elsevier.

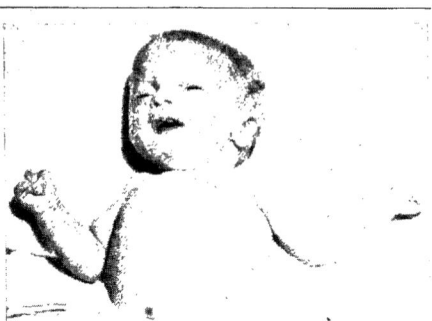

Fig. 1—Appearance of the child showing triangular mouth, low-set ears, and position of hands.

A NEW TRISOMIC SYNDROME

J. H. EDWARDS
M.R.C.P.
OF THE POPULATION GENETICS RESEARCH UNIT (MEDICAL RESEARCH COUNCIL), OLD ROAD, HEADINGTON, OXFORD

D. G. HARNDEN
Ph.D. Edin.
OF THE RADIOBIOLOGICAL RESEARCH UNIT (M.R.C.), HARWELL [*]

A. H. CAMERON
M.D. Durh.
CONSULTANT PATHOLOGIST,
CHILDREN'S HOSPITAL, LADYWOOD, BIRMINGHAM

V. MARY CROSSE
O.B.E., M.D. Lond., D.P.H.
CONSULTANT PÆDIATRICIAN, BIRMINGHAM REGIONAL HOSPITAL BOARD

O. H. WOLFF
M.D. Cantab., M.R.C.P., D.C.H.
SENIOR LECTURER IN PÆDIATRICS, UNIVERSITY OF BIRMINGHAM

THERE are 22 pairs of autosomes in man, and it is possible that 22 syndromes of autosomal trisomy exist; some, however, may be incompatible with intra-uterine life. Mongolism is due to trisomy of one of the smallest chromosomes (Lejeune et al. 1959) and, as it seems reasonable to suppose that the larger the extra autosome the worse the disturbance, most other trisomic syndromes might be expected to be at least as severe. In man such aberrations would probably lead to generalised disorders of shape and size and to multiple structural abnormalities rather than to a single localised deformity.

[*] Present address: M.R.C. Group for Research on the General Effects of Radiation, Department of Radiothrapy, Western General Hospital, Crewe Road, Edinburgh, 4.

One such syndrome is, we believe, exemplified by the following case.

Clinical Record

The infant, a girl, was delivered at term by cæsarean section (Mr. G. S. Lester) at Marston Green Maternity Hospital on May 22, 1959.

Her mother and father, aged 31 and 32 respectively, and her brother, aged 6, are healthy. Between the births of her two children the mother had two miscarriages. During this pregnancy hydramnios was suspected, and she had mild toxæmia. Cæsarean section was carried out because of disproportion.

The birth weight was 5 lb. 1 oz., length 17 in., head circumference 13$^{1}/_{2}$ in. Some unusual features were immediately obvious: an odd-shaped head with wide occipitoparietal and narrow frontal diameters; a broad and flat bridge of the nose; low-set ears; a small mouth, inadequate for breast-feeding; webbed neck; hypermobility of the shoulders, so that they could almost be made to meet; and very short big toes with webbing between the 2nd and 3rd.

On the second day a loud systolic murmur was noted, with its maximum intensity in the 4th left intercostal space near the sternum; the femoral pulses were normal. During the first week of life she had several cyanotic attacks and transient jaundice; then her condition improved and after four weeks she was discharged. At six weeks she again became jaundiced. She was admitted to Little Bromwich Hospital and, after investigation had shown the jaundice to be obstructive, was transferred to the Children's Hospital, Birmingham.

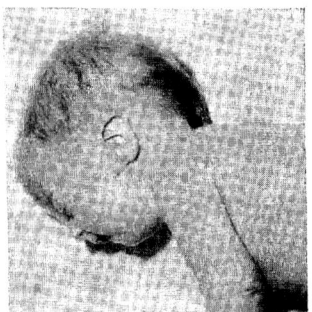

Fig. 2—Photograph showing shape of head and of ears.

She was then 9 weeks old and weighed only 6 lb. 3 oz.; length 19$^{1}/_{2}$ in.; head circumference 14$^{1}/_{2}$ in. Some other features had now become obvious (figs. 1 and 2)—a triangular mouth, a receding chin, and a wide-open metopic suture. The ears were long and oblique, the main axis lying downwards, forwards, and inwards. All the components of the ear were present, but their relative proportions were abnormal. In particular there was an abnormally great distance between the crus antihelicis and the helix margin, giving the ear a pixie-look. The chest was shield-like with nipples almost in the anterior axillary line; the fingers and toes were short and stubby with short irregular and flat nails; the fingers were tightly clenched though they could be straightened. The carrying-angle of the elbow was normal and the external genitalia were normal. The liver was enlarged. The systolic blood-pressure was 130 mm. Hg. Mental development was retarded: she did not smile, focus, or listen to voices.

Investigations allowed us to rule out the following conditions as causes of the jaundice: galactosæmia, syphilis, toxoplasmosis, Weil's disease, glandular fever, and hæmolytic disease of the newborn.

Summary of Other Investigations

Urine.—No inclusion bodies seen.
Blood.—White cells: drumsticks present. Group A Rh+. Haptoglobins type 2 : 2; transferrins type C.
Buccal smear.—Chromatin-positive.
X-rays (R. Astley).—On screening of the *chest*, the heart was seen to be moderately enlarged with a rather globular shape; some pulmonary plethora. The descending aorta had an uneven contour in its upper part. The appearances suggested left-to-right shunt. *Skeleton:* bone age normal; skull, clavicles, and cervical spine normal; thorax dome-shaped. *Pelvis:* iliac index normal according to the data of Caffey and Ross (1956).

Progress

The jaundice persisted, uninfluenced by a ten-day course of prednisolone. After a needle biopsy of the liver (see below) an exploratory operation was performed by Mr. Arnold Gourevitch, on Sept. 21: the liver was greatly enlarged, the gallbladder was small and contained bile; the right, left, and common hepatic ducts and the common bile duct were present but very small. The ovaries, fallopian tubes, and uterus were normal. Cholecyst-jejunostomy was carried out and biopsy specimens obtained from the liver and right ovary. After operation the jaundice deepened and bleeding occurred from the wound and into the bowel. Despite repeated blood-transfusions the infant died on Oct. 2.

Pathological Findings

Liver Biopsy

The needle biopsy and that taken at laparotomy showed the appearances of giant-cell hepatitis (fig. 3). The portal tracts and peripheral parts of the lobules were normal. The central parts contained numerous bile thrombi and showed disorganisation with multinucleated giant cells and interstitial fibrosis.

Biopsy of Right Ovary

Normal ovarian tissue with a normal complement of immature and maturing follicles.

Necropsy

The external features have already been described.
Hæmorrhage from the cholecystojejunal anastomosis filled

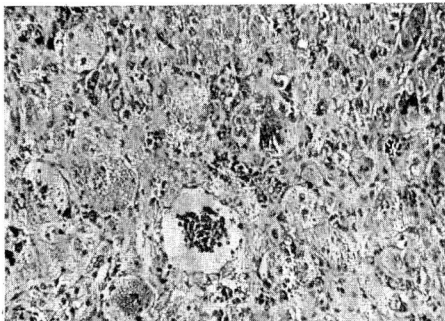

Fig. 3—Low-power view of centrilobular zones of liver showing many multinucleated giant liver cells. (Hæmatoxylin and eosin.)

the small and large intestine. The mesentery was unfixed. The *liver* was firm, was considerably enlarged, and had a smooth external surface. The cut surface was green and the lobular pattern accentuated. The intrahepatic bileducts were not dilated. The common bileduct was small in calibre but patent, and its mucosa was bilestained. The hepatic and cystic ducts were not dissected out, because of adhesions. The *heart* weighed 41·5 g. Both ventricles were hypertrophied and dilated and there was a high ventricular septal defect, 0·6 cm. in diameter. The *ductus arteriosus* admitted the passage of a probe 2 mm. in diameter. The *left ovary* and the remaining part of the *right ovary* were normal. The *spleen* was moderately enlarged, firm, and showed congestion of the pulp and slightly increased trabecular markings. The *thymus* was markedly involuted and lay behind the innominate vein. The *skull* showed a patent metopic suture continuous with a wide anterior fontanelle. The anterior third of the *falx cerebri* was narrow and did not extend for the normal distance into the longitudinal cerebral fissure. The brain was slightly underweight (456 g.). In both frontal lobes and the left parietal lobe the gyri were unusually prominent but not irregular; the sulci were wide and there was a moderate excess of subarachnoid fluid. The arachnoid, like the falx, did not penetrate deeply between the two frontal lobes but there was no abnormal continuity of cerebral tissue crossing the midline. Slicing after fixation showed no further abnormality. There was slight but definite thoracic scoliosis. The *sternum* was thinner than normal and wide, and contained two long thin centres of ossification.

Histological examination of a portion of liver from near the hilum showed normal portal ducts and bileducts. There was fibrosis in the centrilobular zones with considerable loss and disorganisation of liver cells. Many were arranged in acinar groups. Cells with three or four nuclei were common but there were few giant cells. Three small extrahepatic bile ducts were seen in the hilar tissue. One contained granular bile pigment.

Cytological Examinations

Tissue specimens were removed during postmortem examination three hours after death. These were collected in sterile Glaxo medium 199 and set up in tissue cultures twenty hours later using the method described by Harnden (1960). Samples of ovary and kidney proved to be infected and these were discarded but successful cultures were established from skin and muscle taken from the scapular region. Cytological studies on the muscle culture were carried out thirty days after the tissue was established in culture and on the skin culture after twenty-one days. These showed that the great majority of cells from both cultures contained 47 chromosomes:

Tissue	Chromosome counts						Sex chromosomes
	<45	45	46	47	48	>48	
Skin	..	1	..	22	3	..	XX
Muscle	1	..	2	35	..	..	XX

Counts other than 47 can probably all be attributed to technical errors such as breakage of cells during preparation. Six cells from the muscle culture and five cells from the skin culture were carefully analysed and all showed that there were 5 chromosomes resembling pairs 17 and 18 (Ford et. al. 1958) of the normal complement in size and position of the centromere (fig. 4). On the basis of examination of the analysed cells and of other cells not completely analysed, it was decided that there were 3 chromosomes indistinguishable from the no. 17 chromosomes of the normal complement. The most likely explanation for the defect is that the extra chromosome has arisen as a result of non-disjunction during gametogenesis in one of the parents, and that the patient was therefore trisomic for the no. 17 chromosome.

Discussion

The main abnormalities were an odd-shaped skull, low-set and malformed ears, a triangular mouth with receding chin, webbing of the neck, a shield-like chest, short stubby fingers and toes with short nails, webbing of toes, ventricular septal defect, mental retardation, and neonatal hepatitis. The presence of the hepatitis was probably fortuitous but the constellation of other abnormalities is consistent with the type of disorder to be expected from autosomal trisomy, and it seems likely that it may denote a clinical syndrome as specific as mongolism.

The nomenclature and classification of syndromes in which multiple defects occur in association with webbing of the neck are confused, but it now seems possible that chromosomal studies of such cases will lead to clarification. However, until agreement has been reached on a system of numbering or lettering the human chromosomes it would be premature to advance a new name for the syndrome described in this paper.

In the meantime, we suggest that, of the several eponymous syndromes described, only the term "Turner's

Fig. 4—Chromosomes of cell (× approx. 800).

syndrome" (Turner 1938) should be retained, and that it should refer to females with varying degrees of gonadal dysgenesis and some of several associated abnormalities. Of these, short stature, neck webbing, and cubitus valgus occur most commonly; lymphœdema of the extremities, congenital heart-disease, peculiar ears, renal anomalies, a shield-like chest, osteoporosis, and osteochondrosis of the spine, and many others, have also been described. Most cases are chromatin-negative. The majority so far examined have been shown to have an XO sex chromosomal constitution (Ford et al. 1959, Fraccaro et al. 1959) or are XO/XX mosaics (Ford 1959). Our patient with normal ovaries, normal sex chromatin, normal sex chromosomes, and an abnormal set of autosomes cannot be regarded as an example of this syndrome.

Some features of Turner's syndrome were included by Ullrich (1936) in his definition of the Bonnevie-Ullrich syndrome, and our case might be regarded as an example of the latter condition. However, in later writings Ullrich (1938, 1949) has unfortunately so widened the scope of this syndrome, in which he now includes such

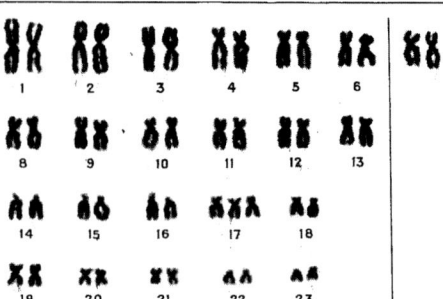

Fig. 5—Chromosomes arranged in pairs.

diverse abnormalities as congenital amputations, hygroma, popliteal webbing, and agenesis of the cranial nerve nuclei, that the term lacks precision. Another disadvantage inherent in its use is the eponymic implication that these conditions are analogous in their genesis to the familial anencephaly in the mouse which has been extensively studied by Bonnevie (1934). Further in the English-speaking countries the term Bonnevie-Ullrich syndrome is widely used as a synonym for Turner's syndrome (e.g., Nelson 1959).

A few cases similar to the present one have been reported.

Bizarro (1938) described a mentally defective boy with widespread abnormalities including webbing of the neck and low ears. James (1952) described a mentally defective boy of 2 years, with low ears, webbing of the neck, slight webbing of the middle and ring fingers, inward curving of the little finger, and a loud systolic murmur. The photograph shows short big toes. Rossi and Caflisch (1951), classifying the various types of webbing, described, as an example of "status Ullrich bilateralis", a 12-year-old girl of short stature with webbing of neck, low-set malformed ears, a shield-like chest, and widely set eyes. Puberal development was normal.

These three cases and ours do not differ more from each other than would four arbitrarily selected mongols, and the possibility of chromosomal identity seems possible. For further clarification it seems essential that other similar syndromes should, when possible, be described in association with chromosomal studies.

It remains to be explained how some of the multiple abnormalities associated with an additional autosome in our case may also occur in association with a missing sex chromosome in Turner's syndrome.

Summary

A female infant presenting with a peculiar facies, webbing of the neck, congenital heart-disease, neonatal hepatitis, and many minor abnormalities was found on postmortem chromosomal study to have an extra chromosome apparently identical to the 17th pair (Ford's nomenclature). This is the second condition of autosomal trisomy to be reported in man.

We are grateful to Dr. C. E. Ford and Prof. D. H. Hubble for much help with the preparation of this report; to Dr. Elizabeth Robson for defining the haptoglobin and transferrin types; to Dr. Sylvia Lawler for attempting to determine the blood-groups from an old specimen of blood taken before extensive transfusion; to Miss C. M. Scammell for technical assistance; to Mr. D. B. Peakman for the photographs.

References overleaf

MULTIPLE CONGENITAL ANOMALY CAUSED BY AN EXTRA AUTOSOME

KLAUS PATAU
Ph.D. Berlin

DAVID W. SMITH
M.D. Baltimore

EEVA THERMAN
Ph.D. Helsinki

STANLEY L. INHORN
M.D. New York

HANS P. WAGNER
M.D. Berne

From the Departments of Pathology and Pediatrics, Medical School, University of Wisconsin, Madison, Wisconsin

SINCE the introduction of new methods which made possible the reliable determination of the chromosome number in man and the identification of individual chromosomes or at least groups of chromosomes, it has become evident that the presence of the normal complement of exactly 22 pairs of autosomes and 2 sex chromosomes is essential for normal development.

Abnormal chromosome numbers are most likely to arise by non-disjunction in the first meiotic division. Should this happen to the sex chromosomes in either parent, fertilisation will result in one of four abnormal sex-chromosome constitutions. Of these possibilities one, YO, would unquestionably be lethal. All of the remaining combinations have been observed. XO, with a total of 45 chromosomes, causes gonadal dysgenesis (Ford, Jones, Polani, de Almeida, and Briggs 1959), and XXY chromatin-positive Klinefelter's syndrome (Jacobs and Strong 1959, Ford, Jones, Miller, Mittwoch, Penrose, Ridler, and Shapiro 1959). XXX results in disturbances (Jacobs, Baikie, Court Brown, MacGregor, Maclean, and Harnden 1959) that seem to indicate merely an œstrogen deficiency. On genetical grounds it was not to be expected that the addition of an autosome to the normal complement would have a similarly restricted effect. Only one type of autosomal trisomic has been reported to date, and although the extra chromosome is one of the two smallest autosomes of the haploid set, its presence in triplicate results in mongolism (Jacobs, Baikie, Court Brown, and Strong 1959, Lejeune, Turpin, and Gautier 1959).

It was to be expected that other autosomal trisomics, if they should be at all viable, would also display multiple congenital disturbances. During a search among infants afflicted with such anomalies we have recently found two clinically similar cases, a boy and an unrelated girl, each with 47 chromosomes. The syndrome is quite distinct from mongolism and includes anomalies of ears, hands, and feet, a small mandible, apparent mental retardation, spasticity, and a congenital heart defect. The heart defect proved fatal at 2 and $2^{1}/_{2}$ months of age. The cytological analysis is still in progress; it seems that the extra chromosome belongs to the E group (in our classification as proposed in fig. 2). The present communication concerns a third type of autosomal trisomic which is distinct from mongolism as well as from the above-mentioned syndrome.

Clinical Data

The patient, a full-term female infant, was born in January, 1959, and is still alive (February, 1960).

This was her mother's first pregnancy. Both parents are Caucasian; they were 25 years of age and in good health at the time of conception. A survey of the parental family trees disclosed 3 first cousins, 7 aunts and uncles, and 31 great-aunts and great-uncles with no reported anomalies. The parents had 89 first cousins, of whom 1 had a lumbar meningocele. One of the patient's 88 second cousins has a clubfoot deformity, another a congenital shortness of one leg. The grandparents are said to be normal. Thus among 224 relatives of three generations only three anomalies were found, none similar to those observed in the patient.

At 6 months' gestation the mother had an influenza-like illness. She received no X-ray exposure during the pregnancy and experienced no bleeding. At birth, the patient weighed 6 lb. 5 oz. and was 19 in. in length. Intermittent cyanosis was noted in the first 3 postnatal hours and oxygen was administered in an incubator during the first 3 days. No cyanosis has been observed since.

The infant was first seen at the age of 1 month by one of us (D. W. S.). She appeared to be a well-nourished infant with the following anomalies: apparent anophthalmia, hare lip, cleft palate, and polydactyly of the left foot (fig. 1). She weighed 7 lb. 13 oz., was 21·0 in.

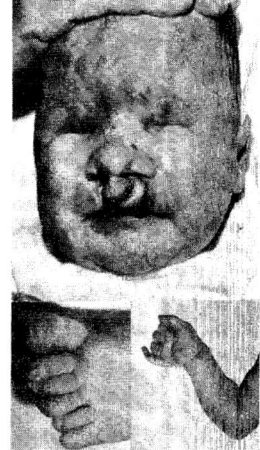

Fig. 1—The patient.

long, and had a head circumference of 14·25 in., and a chest circumference of 14·0 in., all of which are within the normal range. The appearance of the external genitalia was that of a normal female infant.

The cranium was normal in size and contour and could not be transilluminated. Poor ossification of the cranial bones with wide membranous gaps between them was revealed by X-rays. When the eyelids, which seemed to be normal, were pried open no organised ocular tissue could be seen or palpated. The condition of the posterior orbit remains unknown as no biopsy was taken. The bilateral hare-lip deformity was worse on the right side. The cleft in the palate was complete. The left ear was smaller than the right. In the 6th toe of the left foot X-rays disclosed normal phalanges based on a fully developed 6th metatarsal. Both thumbs were maintained in a flexed position. Upon their passive extension two small " clicks " were palpable at the metacarpal-phalangeal joint, an anomaly sometimes called " trigger thumb ".

A grade-III rough systolic murmur was audible to the left of the sternum with maximal intensity over the 2nd to 3rd intercostal spaces. The blood-pressure in the right arm was 90 mm. Hg by flush technique and the pulse-rate was 120. Chest X-rays and fluoroscopy revealed a globular heart of

---

DR. EDWARDS & OTHERS: REFERENCES

Bizarro, A. H. (1938) *Lancet*, ii, 828.
Bonnevie, K. (1934) *J. exp. Zool.* **67**, 443.
Caffey, J., Ross, S. (1956) *Pediatrics*, **17**, 642.
Ford, C. E. (1959) King's College Hospital Symposium, London.
— Jacobs, P. A., Lajtha, L. G. (1958) *Nature, Lond.* **181**, 1565.
— Jones, K. W., Polani, P. E., de Almeida, J. C., Briggs, J. H. (1959), *Lancet*, i, 711.
Fraccaro, M., Kaijser, K., Lindsten, J. (1959) *ibid.* p. 886.
Harnden, D. G. (1960) *Brit. J. exp. Path.* **41**, 31.
James, T. (1952) *Edinb. med. J.* **59**, 344.
Lejeune, J., Gautier, M., Turpin, R. (1959) *C. R. Acad. Sci., Paris*, **248**, 602.
Nelson, W. E. (1959) Text Book of Pædiatrics; p. 1198. Philadelphia and London.
Rossi, E., Caflisch, A. (1951) *Helv. pædiat. Acta.* **6**, 119.
Turner, H. H. (1938) *Endocrinology*, **23**, 566.
Ullrich, O. (1936) *Handb. Neurol.* **16**, 139.
— (1938) *Klin. Wschr.* **17**, 185.
— (1949) *Amer. J. hum. Genet.* **1**, 179.

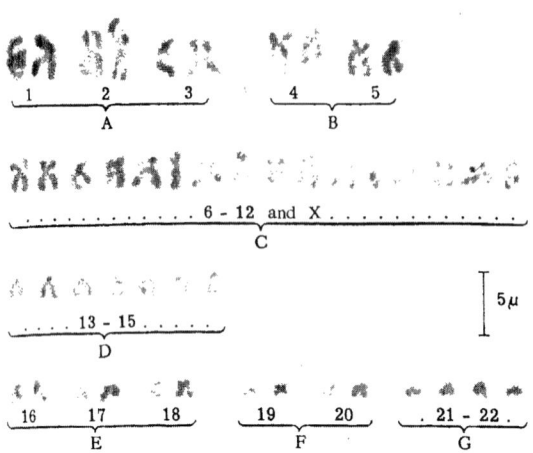

Fig. 2—The 47 chromosomes of a cell from the bone-marrow of the right tibia. The D group contains 7 instead of, as normal, 6 autosomes.

normal size with the main mass of the heart projecting in a convex manner to the right. The aorta was seen to descend on the left. Pulmonary vascularity was increased. The electrocardiogram was interpreted as showing a vertical position with marked right axis deviation. The clinical impression was a rotational anomaly with intraventricular septal defect.

Non-elevated simple capillary hæmangiomata were present on the nasal bridge, both upper eyelids, the posterior neck, and scattered small areas on the forehead and lower back. A neurological study showed that the infant moved all extremities with fair muscular tone. The deep tendon reflexes were normal. A fair Moro reflex was elicited by sudden movement, but no such reaction was provoked by a loud sound. As the infant showed no response to sound she was considered to be deaf. In a subsequent evaluation at 5 months of age it was estimated that she performed at the developmental level of a 1-month-old. Since the age of 3 months she has had frequent brief seizures of myoclonic type.

Laboratory studies of urine and blood yielded normal values, as did measurements of the concentrations in the serum

CHROMOSOME COUNTS IN BONE-MARROW MITOSES

Specimen from	Chromosome number						Number of mitoses
	44	45	46	47	48	49	
Right tibia	..	..	..	4	..	1	5
Left ,,	..	..	1	6	..	..	7
Right ,,	..	..	2	37	..	..	39
Total	..	..	3	47	..	1	51

of CO_2, chloride, sodium, potassium, calcium, phosphorus, serum alkaline phosphatase, total protein, and fasting blood-sugar.

Particular attention was paid to the question whether the patient might be an atypical mongoloid. She does have simian creases in the palms of both hands, but in every other aspect of the physical evaluation she did not appear mongoloid. X-ray photographs of the pelvis were normal for her age. She lacked the hypotonia usually evident in an infant mongoloid. In addition, the skin patterns of the hands and feet render mongolism very unlikely. We are obliged to Dr. Irene A. Uchida for a detailed analysis of these patterns. Their overall logarithmic index is −1·81, a value which is smaller than that found by Walker (1957) in all but 2 of 150 investigated mongoloids and which is well within the range of non-mongoloids. This index, in conjunction with the clinical picture, justifies the conclusion that the patient is not a mongoloid. It will be seen that the cytological analysis leads to the same conclusion.

Cytological Observations

The chromosomes were studied in cells from bone-marrow cultures. The culture technique followed was essentially that recommended by Ford, Jacobs, and Lajtha (1958).

Marrow specimens were obtained by aspiration first from the right tibia, later from the left tibia, and still later again from the right tibia. In the case of the first two specimens only the buffy layer of the centrifuged marrow was cultured in saline with AB serum from a blood-bank. The third specimen, without centrifugation, was put into a culture medium (about 0·25 ml. marrow per 4·0 ml. medium) consisting of equal parts of Hanks' balanced saline (with antibiotics, adjusted to pH 7·4) and fat-free serum collected from an AB donor after fasting. The first two specimens yielded only a few countable mitoses; the third produced an abundance. Observations on bone-marrow from other individuals also suggest the last-described technique to be

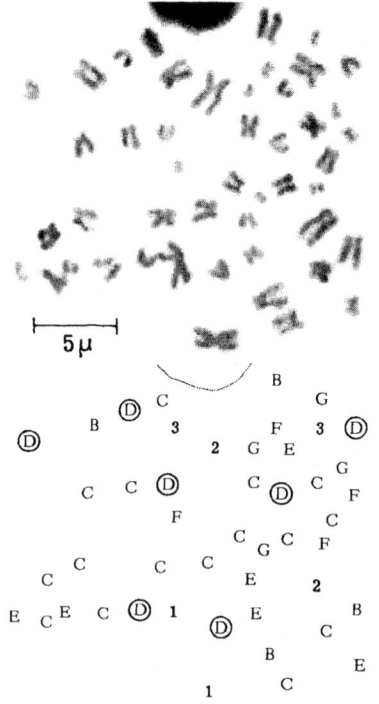

Fig. 3—A cell with 47 chromosomes, including 7 of the D group, from the bone-marrow of the left tibia. The letters designate chromosome groups as in fig. 2.

superior; but it, too, often fails to provide more than a few usable mitoses.

The cultures, 4 ml. aliquots in 25 ml. screw-cap tubes, were kept for approximately 24 hours at room temperature, then for 5 hours in an incubator at 37°C, and after the addition of colchicine for 2 more hours in the incubator. The colchicine concentration in the culture medium was 0·0001%. Sodium citrate was used as by the above authors. After fixation in 1:3 acetic alcohol, the material was either left in 75% alcohol in a refrigerator for future use or stained immediately by the usual Feulgen procedure. The squashing was done in 45% acetic acid. Thereafter the slides were either made semipermanent by surrounding the coverslips with Kroenig cement or mounted in 'Euparal'. To that end the preparations were frozen on dry ice, the coverslips pried off, and the slides passed through 95% into absolute alcohol.

Semipermanent slides have the advantage that microscopically controlled additional squashing can be applied. This often renders the chromosomes countable in a cell that was not suitable before. Even more important, sufficiently severe squashing tends to spread the chromatids of the smaller chromosomes in a horizontal plane, an almost necessary condition if the latter are to be reliably classified. It matters little that squashing, when carried that far, may distort some of the larger chromosomes (see fig. 2).

Chromosomes were counted for the record only when their arrangement suggested that none had been lost by oversquashing. Absolute assurance on this point is impossible; indeed, in one instance of repeated squashing we observed a chromosome being expelled from a cell that continued to have an unbroken appearance. However, such cases are evidently rare. The outlay of the chromosomes of each cell judged to be acceptable in the above sense was first sketched. The decision to count them for the record was made by two of us (K. P. and E. T.) who had both to be satisfied that the sketch contained no errors. We refrained from any attempt to ascertain the chromosome number before this decision. Thereafter the count obtained from the drawing was not open to further revision. It is obviously impossible to estimate, without a deliberate count, the number of some 46 scattered chromosomes with any precision. Therefore, the above procedure will ensure that, within the narrow range of chromosome numbers which are of any interest in the present context, subconscious bias can affect neither the selection of cells to be counted nor the counts themselves. A very few cells with less, mostly much less, than 44 chromosomes were discarded, as were polyploid mitoses which seem to occur in every bone-marrow.

It is evident from the table that the basic chromosome number of the patient is 47. As this number predominates in the marrow of both tibias the possibility of mosaicism can be dismissed. The few cases with numbers other than 47 may have come about in part by loss of chromosomes during preparation, in part by counting errors; but we would not rule out the possibility of having encountered an occasional cell with an atypical chromosome complement.

There was no lack of cells in which all chromosomes or major groups of them could be analysed. In all suitable cells with 47 chromosomes an apparently entirely normal female complement was found, including 4 small chromosomes (group G in fig. 2) and 16 of the group to which the X chromosomes belong (group C in fig. 2). The presence of two X chromosomes was further borne out by buccal smears, fixed and stained with acetic orcein, in which 25 out of 100 nuclei were chromatin-positive. The extra chromosome belongs to the group of medium-sized acrocentric autosomes (D in fig. 2). In at least five cells all 7 D chromosomes were about as clear as in fig. 2, each showing the very small short arm characteristic for this group. There were, of course, still other cells—e.g., that of fig. 3, in which all 7 D chromosomes could be identified.

We cannot say to which of the three pairs of the D group the extra chromosome belongs, the reason being that we cannot identify the individual pairs. It will be pointed out elsewhere that the "pairing-off" of chromosomes by size in the case of this, as of certain other groups, is almost meaningless because it neglects the not at all negligible random variation of the apparent chromosome length. One of the D pairs has a minute satellite (Tjio and Puck 1958, Chu and Giles 1959). We have seen it, but too infrequently—even in orcein slides made for this purpose—to have much hope that the presence or absence of a satellite at the extra chromosome could be established with the present technique.

Discussion

It will, we presume, not be doubted that the observed abnormality of the chromosome complement is the cause of the clinical anomalies; but the question could be raised whether the extra chromosome might not be a translocation chromosome which merely happens to have the size and shape of a D chromosome. If so, the chromosome piece actually duplicated could be considerably shorter than a whole D chromosome. Suffice it to say that for various reasons we consider this unlikely in the present case and that we are confident future findings will confirm the extra chromosome to be a genuine D chromosome. If it is, the present combination of a specific cytological situation with a certain pattern of anomalies is bound to be encountered again. In independent cases of translocation a similar clinical pattern might reoccur but would not consistently coincide with the presence of an extra chromosome of the D type.

Trisomy no doubt always has its origin in non-disjunction. The latter represents, if the original chromosome complement is structurally homozygous, a meiotic or (more rarely) a mitotic malfunction. Such an accident can involve any chromosome, even though some chromosomes may be much more prone to it than others, depending on their length and centromere position. We suspect that non-disjunction occurs occasionally in the healthiest of tissues. The frequency of this as of any malfunction is, however, apt to be influenced by genetical and environmental factors. One need not be surprised that there is a correlation between the incidence of mongolism, now understood to result from non-disjunction, and the age of the mother, which, after all, is to the oocyte an environmental factor. This strong positive correlation is not likely to be restricted to one particular chromosome. In consequence, we should expect that the frequency of trisomics other than mongoloids will also be found to increase with the age of the mother. This appears to be the case. Out of four such trisomics observed so far only the present patient has a young mother. The mothers of the two autosomal trisomics of another type that had been mentioned in the introduction were both about 46 years of age at the time of conception. The mother of the XXX female reported by Jacobs, Baikie, Court Brown, MacGregor, Maclean, and Harnden (1959) was 41 years of age at conception. These are not likely to be mere coincidences.

Of the 22 conceivably viable types of autosomal trisomics three have been found so far. How many more are viable is not known, but it seems certain that not all of them are. As each of the viable ones will probably show a clinically distinguishable syndrome, we may expect that an ætiologically unique group of a limited number of "autosomal trisomy syndromes" will become established. To the human geneticist these will be of continuing interest; he will in particular look forward to cases in which the presence of known genes in the parents can be related to peculiarities in the trisomic child. It seems likely that every component anomaly of a trisomy syndrome reflects, by way of a dose effect of gene action, the presence in the respective chromosomes of at least one gene locus that plays a prominent role in the normal development of the afflicted organ; indeed, we suspect that many of the trisomy anomalies can also be produced individually in euploid persons by the heterozygous or homozygous presence of a suitable mutated allele of the responsible gene. Polydactyly and hare lip with cleft palate in the present patient may be cases in point. These also suggest that trisomy may become instrumental in establishing for the first time autosomal linkage groups in man.

Summary

The patient, a girl infant, has 47 chromosomes, the extra chromosome being one of the medium-sized acrocentric autosomes.

The presence of this extra chromosome is regarded as the cause of the following observed combination of congenital anomalies:

Cerebral defect	Simian creases
Apparent anophthalmia	"Trigger thumbs"
Cleft palate	Polydactyly
Hare lip	Capillary hæmangiomata
	Heart defect

We wish to thank Mr. H. Montague and Mr. J. Tiedt from the photographic laboratory for technical assistance.

The work has been supported in part by a grant from the U.S. Public Health Service (C-3313) and by an institutional grant from the American Cancer Society, Inc.

Addendum

Since this report was written, another patient, an unrelated female infant, has been found to combine a similar set of congenital anomalies with possession of an extra D chromosome which no doubt is the same as the one in the present case.

REFERENCES

Chu, H. Y., Giles, N. H. (1959) *Amer. J. hum. Gen.* **11**, 63.
Ford, C. E., Jacobs, P. A., Lajtha, L. G. (1958) *Nature, Lond.* **181**, 1565.
— Jones, K. W., Miller, O. J., Mittwoch, U., Penrose, L. S., Ridler, M., Shapiro, A. (1959) *Lancet*, i, 709.
— — Polani, P. E., de Almeida, J. C., Briggs, J. H. (1959) *ibid.* p, 711.
Jacobs, P. A., Baikie, A. G., Court Brown, W. M., MacGregor, T. N., Maclean, N., Harnden, D. G. (1959) *ibid.* ii, 423.
— — — Strong, J. A. (1959) *ibid.* i, 710.
— Strong, J. A. (1959) *Nature, Lond.* **183**, 302.
Lejeune, J., Turpin, R., Gautier, M. (1959) *Ann. Génétique*, **1**, 41.
Tjio, J. H., Puck, T. T. (1958) *Proc. nat. Acad. Sci., Wash.* **44**, 1229.
Walker, N. F. (1957) *J. Pediat.* **50**, 19.

CHAPTER 6

Chromosomes and leukaemia

THE IMPORTANCE OF CHROMOSOMES in the initiation of cancer has been recognised since the time of Boveri's 1914 monograph[1] and chromosome abnormalities were recognised in cancer cells from at least the beginning of the 20th century.[2] Indeed, the occurrence of numerous, easily visible and often abnormal mitoses became one of the cytological hallmarks of malignancy. The later development of techniques of tissue and cell culture, and the experimental use of tumours growing as monolayers such as ascites tumour cells, gave a powerful stimulus to cancer cytogenetics in the 1950s, especially when combined with hypotonic chromosome spreading. Cancer chromosomes were thus in some ways easier to study (though not necessarily to interpret) than normal human chromosomes; it should be remembered (Chapter 2) that Levan's main reason for wanting to establish the normal human chromosome number was to provide an accurate control for comparison with his tumour studies.

By the late 1950s there were a number of groups worldwide involved in cancer cytogenetic research, including those of Theodore Puck, TC Hsu, Ernest Chu and Theodore Hauschka in America, Albert Levan and PC Koller in Europe, while in Japan the work of Sajiro Makino and Akio Awa had been given special impetus by the tragic effects of the atomic explosions. The reviews of Levan[3] and Hauschka[4] summarise the state of knowledge at this time and a clear idea of the detailed nature of the work can be seen in the meticulous drawings of Levan in his 1956 account of his collaborative studies at Sloan-Kettering Institute, New York[5] (Figure 6-1).

Despite this extensive activity, no specific chromosome patterns could be identified in human cancer cells, and it was only following Tjio and Levan's definition of the normal human karyotype[6] that the field of cancer genetics could develop in a systematic manner. Even then, it remained unclear whether the changes seen were primary and causative, or whether they reflected secondary damage to chromosomes by viral and other external factors. Most of the key developments came after the period covered by this book, being

Fig. 6-1 Camera lucida drawings of cancer chromosomes. From Levan, 1956.[5] (Courtesy of *Cancer*.)

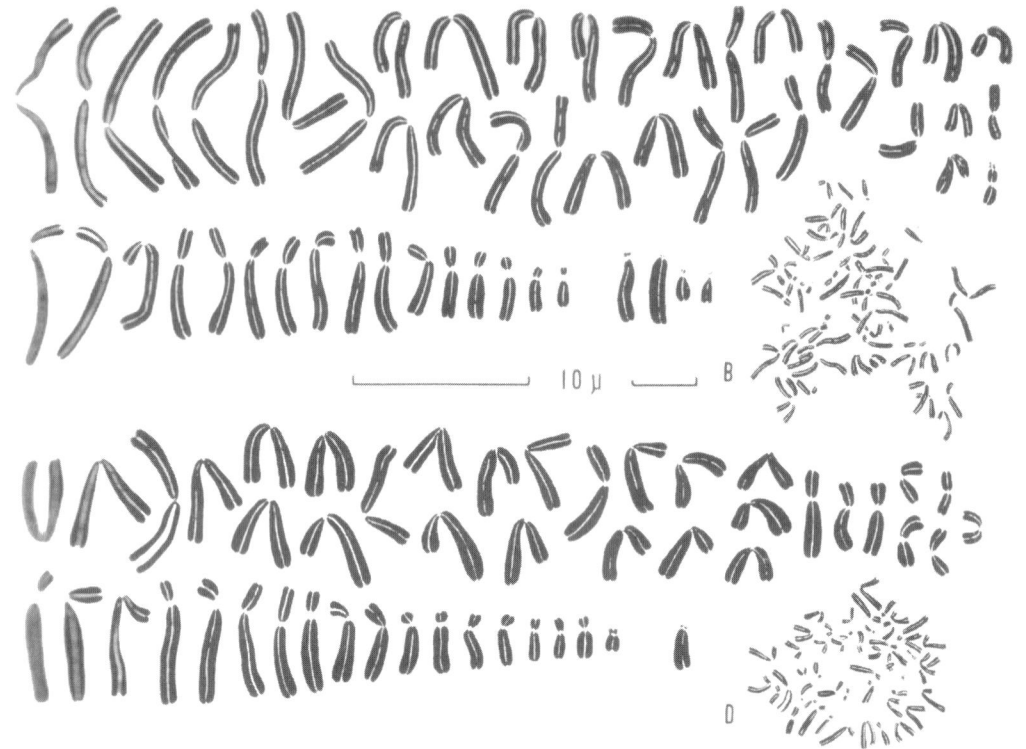

critically dependent on the analysis of detailed morphology made possible by chromosome banding techniques from 1970 onwards.

The single exception to this is provided by the leukaemias, notably chronic myeloid leukaemia, where the initial discoveries belong to the same years, 1959–1960, that also saw the first developments in recognising constitutional chromosome abnormalities, sharing the same roots in terms of technology and to some extent the same groups of investigators. This chapter therefore focuses on leukaemia, rather than on solid tumours.

In tracing the origins of leukaemia cytogenetics, the remarkable work of the Russian workers in the 1930s, already mentioned in Chapter 1, makes a suitable starting point. Andres and Shiwago in 1933[7] reported detailed results in two cases of myeloid leukaemia, one acute, the other seemingly chronic; they used cultured blood and were able to compare their findings with normal controls by using the method of blood culture already described by their colleagues in 1931[8] and discussed in more detail in the next chapter, along with hypotonic treatment. They did not find any specific abnormality, but in their techniques they were 20 years ahead of investigators elsewhere.

It was not until the later 1950s that techniques of bone marrow culture were developed that would allow detailed chromosome analysis in leukaemia. This was the result of a collaboration between three British research groups. The Edinburgh 'Clinical Effects of Radiation Research Unit' of the Medical Research Council (MRC), under Michael Court Brown, was involved in a major epidemiological and aetiological study of radiation-induced leukaemias; the closely linked MRC Radiobiology Unit at Harwell, England (director John Loutit), had a cytogenetics group led by Charles Ford, whose wider work has already been described in Chapters 2 and 4; finally, the expertise in bone marrow culture was provided by Laszlo Lajtha, based in nearby Oxford University. The Harwell unit had been strengthened by the move to there from Edinburgh a few years previously of the mouse genetics group, allowing radiation and cytogenetic studies to be developed on mammals and making use of the atomic reactor facilities at Harwell.

The background to the setting up of the Edinburgh MRC unit, initially titled the 'Group for Research into the General Effects of Radiation' is well described in an unpublished and unsigned progress report

W MICHAEL COURT BROWN (1918–1969)

Trained initially in radiotherapy but turning to the epidemiological basis of cancer, in particular leukaemias, Michael Court Brown joined the Medical Research Council (MRC) and in 1956 was appointed head of the new Clinical Effects of Radiation Unit based at Western General, Edinburgh. He was responsible for initiating cytogenetic techniques in the unit, with the appointment of Patricia Jacobs, David Harnden and others; the combination of epidemiological, clinical and cytogenetic approaches proved especially fruitful, not just in the leukaemia and radiation fields but in human cytogenetics generally. His wider work included the discovery, with Richard Doll, that medical irradiation for ankylosing spondylitis could cause leukaemia. Described by colleagues as frank-speaking but extremely supportive, his early death in 1969 was a major loss to the unit and to science generally. (Photo courtesy of Professor David Harnden.)

written in 1960 and forming part of the unit's valuable archive. Written almost certainly by Michael Court Brown, it describes the unit's setting up in 1956 in response to widespread concern about nuclear fallout and increase in leukaemia frequency, with a tripartite research programme of epidemiological analysis, cytogenetic work and clinical studies of leukaemia patients.

Patricia Jacobs, then a young graduate scientist, was appointed to the Edinburgh unit in October 1957 to develop chromosome techniques in relation to the leukaemia studies, and was sent to Harwell and Oxford to learn and combine these various techniques; her memories from this time have already been quoted in earlier chapters and other background information is given in the MRC progress report mentioned above. The resulting paper, published in *Nature* in 1958,[9] laid secure foundations for further chromosome studies of leukaemias in Edinburgh, but neither it nor the subsequent paper of the Edinburgh workers in 1959, giving the results on 12 cases, including four of acute leukaemia and five of chronic myeloid leukaemia,[10] found any specific abnormalities.

The geographical focus of the story now shifts markedly, for while most of the cytogenetic discoveries presented so far in this book were European in their origin, the initial discoveries in leukaemia cytogenetics were American, based on a remarkable collaboration in Philadelphia between a basic cytogeneticist, David Hungerford,

Fig. 6-2 Peter Nowell (left) and David Hungerford, with a cytogenetic background. (Courtesy of Fox Chase Cancer Center Archives.)

and a clinically trained pathologist, Peter Nowell (Figure 6-2). Nowell has given accounts of their early joint work and discoveries,[11,12] and the success of the collaboration seems to have been based, as is often the case, on the differences in their respective contributions. Hungerford was a young cytogeneticist who had come from basic *Drosophila* and plant cytogenetics to study human chromosomes at Fox Chase Cancer Center. Nowell had begun leukaemia research at University of Pennsylvania School of Medicine following

PETER NOWELL (born 1928)

Peter Nowell began medical studies at University of Pennsylvania School of Medicine in 1948 and states in his 'personal perspective' that he already knew that he wished to pursue cancer research. A pathology residency stimulated his interest in leukaemias, which was increased by radiation research studies during a period as a US Navy medical officer. On return to Philadelphia in 1956, he made the link with David Hungerford, as described in the text and, like Hungerford, remained in Philadelphia for the rest of his career, becoming head of Haematology and later of the university cancer centre. Returning to research when chromosome banding and molecular analysis were producing a new wave of cancer genetics research, he remains actively involved in the field. He has given two historical accounts of his career,[11,12] as well as a joint description, with Janet Rowley and Alfred Knudsen, of the key early developments in cancer genetics.[21] (Photo courtesy of Dr Peter Nowell, reprinted, with permission, from the *Annual Review of Medicine*, Volume 53 © 2002 by Annual Reviews www.annualreviews.org

work during his military service on radiation pathology.

The collaboration was much more, however, than a fruitful link between clinician and scientist. Nowell's own investigations had led him to use leukaemic peripheral blood cultured with the bean extract phytohemagglutinin to remove red blood cells,[13] paving the way for the more general use of peripheral blood outside the leukaemias (see Chapter 7). He also treated the slides with tap water before staining[12] (he was unaware of TC Hsu's work on hypotonic spreading of chromosomes – or that of Makino and the Russian workers still earlier), so that his preparations showed clear and well spread mitotic divisions. Not having cytogenetic expertise himself, he sought out David Hungerford, who thus had suitable human material to work with from the beginning. Nowell's 'personal perspective' on the origins of the work is well worth quoting.[12]

> *I returned to Philadelphia in 1956 and began some poorly defined studies of leukemia, looking at the growth and differentiation of human leukemic cells in irradiated mice and in vitro.... The leukemic cells were grown on small slides, and, having been partially trained in pathology, I prepared them for examination by rinsing them with tap water before staining them with Giemsa. I had no idea that I was re-enacting the discovery, made earlier by TC Hsu and by others, and often equally serendipitously,*

DAVID A HUNGERFORD (1927–1993)

Born in Brockton, Massachusetts, David Hungerford moved frequently during his early life, since his father was in the US Navy, which he himself joined aged 17, in 1945. Graduating from Temple University, Philadelphia, he initially worked as a technician, then research assistant and junior research fellow in the Fox Chase Institute for Cancer Research, Philadelphia, where his collaborative work with Peter Nowell was done; he remained there and at the University of Pennsylvania School of Medicine for the rest of his career. David Hungerford's research career was halted by the development of multiple sclerosis; one of his last publications, in 1978,[22] is a notable historical review of early studies in human cytogenetics, one of the few such accounts to give full recognition to the early Russian workers in the field. Fox Chase Institute has a full archive of his work, including background documents, and I am most grateful to Beth Lewis, archivist, for providing me with material. (Photo courtesy of Fox Chase Cancer Center Archives.)

that hypotonic treatment could be a major aid to mammalian cytogenetics.

In the meantime, I had not yet met my future collaborator, but I began looking around to see if anyone might be interested in the chromosomes of my leukemic cells. I checked first at the Wistar Institute, across the street, and they referred me to a graduate student at the Institute for Cancer Research in Fox Chase. This was Dave Hungerford, who was trying to do a thesis on human chromosomes with Jack Schulz, but had little material to work with. A natural collaboration resulted. I continued my investigation of leukemic cell differentiation in culture, but also prepared some slides, under the more sophisticated direction of Dave Hungerford, and sent them up to him for karyotypic analysis.

The personalities, as well as the skills of the two workers, may also have been complementary, though one should not perhaps take too seriously the views of the Philadelphia journalist who produced the following press release after their discovery!

Peter C Nowell at 33 is a buoyant, garrulous doctor of medicine. His colleague, David A Hungerford, 34, is an intense, soft-spoken doctor of philosophy in zoology. Working together in Philadelphia, Nowell at the University of Pennsylvania's School of Medicine and Hungerford at the Institute for Cancer Research, they have spent four

years exploring the tiny, mysterious world of human chromosomes.
(Reproduced from Fox Chase Cancer Center Archive, courtesy of Beth Lewis, archivist.)

Nowell and Hungerford's work on chronic myeloid leukaemia provides a landmark in human cytogenetics, marking the transition from simply counting the chromosome number to analysing the detailed morphology of chromosomes. Their first two patients showed, unlike patients with other forms of leukaemia, an abnormally small chromosome consistently present (Figure 6-3), which replaced one of

Fig. 6-3 Metaphase showing Ph^1 chromosome from Nowell and Hungerford's original study. (Courtesy of Dr Peter Nowell and *Annual Review of Medicine*.)

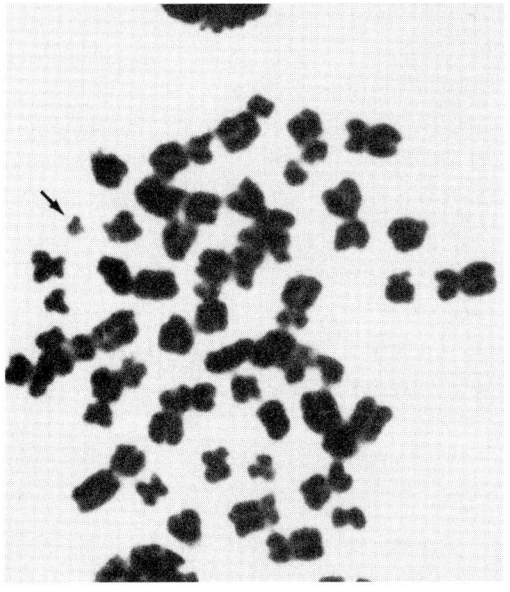

the other small acrocentric chromosomes,[14,15] though the actual chromosome number was normal. Both of these first patients were male so it was initially suggested that the abnormal chromosome might be derived from the Y chromosome, being reported as such in their initial paper (see Addendum).[14] It was only when a larger series was published[16] that it became clear that the chromosome of origin was one of the larger 'G group' pair of chromosomes, now designated 22 to avoid disturbing the classification already in use for 'trisomy 21' in Down's syndrome.

While this work was in progress in Philadelphia, the Edinburgh unit, after their initial normal results, had extended their study and had also begun to use peripheral blood from their leukaemic patients.[17] Again the chromosomes appeared normal in most types of leukaemia, but when they concentrated on patients with chronic myeloid leukaemia they, like Nowell and Hungerford, also found an abnormally small chromosome to be present.[18] They also showed that the chromosomes of cultured skin fibroblasts (using David Harnden's technique) were normal, indicating that the abnormal chromosome in blood and bone marrow must be of somatic rather than constitutional origin.

The influence of the two groups on each other was significant, even though the findings were independent. Nowell remembers[12] that the initial normal findings of the Edinburgh group made him and Hungerford cautious of a causative link when interpreting their own findings, possibly contributing to them ascribing the abnormality to the Y chromosome in their first report. Having the information from Philadelphia available before writing up their own results, the Edinburgh group were able to check patients of both sexes and to conclude that the abnormality was not related to the Y chromosome but to a G group autosome, as also confirmed by Nowell and Hungerford in their second brief report appearing late in 1960 in Science[15] and in their larger series published the following year.[16]

Reading the papers of Nowell and Hungerford 45 years later, one is struck by the high degree of independence of the relatively young workers involved, something already noted in relation to Patricia Jacobs and others in Chapter 3. Hungerford was a graduate student, yet to complete his PhD; Nowell, though clinically trained, was far from experienced in research – as he states, he 'returned to Philadelphia in 1956 and began some poorly defined studies of leukemia'. The key papers do not include any supervisor or departmental head as co-authors and it seems as if the two collaborators were given a free rein to carry their work and ideas in whatever direction they might go. Of course, the high quality research facilities available to them in Philadelphia must have helped, and equally we do not now read of the many unsuccessful first attempts of young researchers half a century ago; but nonetheless, this collaboration reflects well on how the institutions involved chose, nurtured and later gave full credit to their young investigators. As Peter Nowell himself states:[12]

I can only hope that the young people entering the field, and having the newer tools to extend these studies will also find, as I did, an environment that allows inquiry in unexpected directions and easy links to clinical applications. These are the factors that have made, for one investigator, a remarkably satisfying career.

Both Nowell and Court Brown were clinicians with a wide knowledge of leukaemia; the importance of documenting the clinical context in this (and other) research is shown not only by the specificity of the chromosome abnormality to chronic myeloid leukaemia, but in the fact that both groups were able to avoid linking the chromosome involved with that already known to be trisomic in Down's syndrome, where leukaemia is increased in frequency. This was despite the uncertainties at that time of distinguishing chromosomes 21 and 22. Patricia Jacobs is definite that any relationship with Down's had been rejected from the start.

Some people thought that, but I knew it wasn't because I was working with haematologists and they said, 'But that's not the kind of leukaemia Down's get'.

Indeed the Edinburgh group's 1961 paper[18] was able to give results on five Down's patients with leukaemia and found no chromosome abnormality other than the expected trisomy 21.

While the findings of the two groups were independent, the Edinburgh workers freely admitted that Nowell and Hungerford's discovery came first and generously christened the abnormality 'the Philadelphia chromosome', noting in their third (1961) paper:

In this paper the abnormal chromosome, for the sake of brevity, is referred to by the symbol Ph¹.

David Harnden[19] recalls the background to this:

Michael Court Brown set about writing up an account of this 'unusually small, small acrocentric chromosome'. This sounded rather clumsy and Pat Jacobs and I recalled the convention proposed at the Denver meeting, that a chromosome of abnormal morphology should be assigned an 'arbitrary symbol, prefixed by a designation of the laboratory of origin'. So the first abnormal chromosome from Philadelphia, Ph¹ for short, was named in honour of Nowell and Hungerford.

Though this aspect of the Denver nomenclature (see Chapter 7) never caught on, in contrast to the geographical naming of haemoglobins, the term 'Philadelphia chromosome' has remained in the minds of clinicians and medical students ever since, even though many have probably never understood what it actually is.

Further significant progress on the chromosome change in chronic myeloid leukaemia, and in other leukaemias, had to wait for the development of chromosome banding techniques 10 years later; Nowell admits[11] that this intervening decade was a frustrating time and with hindsight it is clear that had the Ph^1 change involved one of the large chromosomes rather than 22, it would not have been detected. It remained uncertain during this time whether the loss of chromosome material from 22 represented a true deletion or whether it was translocated elsewhere, a difference that had general implications for the relationship of cytogenetic abnormalities to neoplasia. Both groups recognised that the relatively small amount of extra chromosomal material involved would not be recognisable with current techniques if translocated onto one of the larger chromosomes. Only with the 1973 finding of Rowley[20] did it become certain that the apparently missing segment was indeed translocated on to chromosome 9q, setting the stage for molecular analysis that would focus cancer genetics on the disruption of specific single genes, rather than on dysfunction of entire chromosomal regions.

David Hungerford's active career in cytogenetics was sadly cut short by the development of multiple sclerosis, as was that of Michael Court Brown by his sudden death in 1969. They and their co-workers had, though, laid the foundations for lasting links between leukaemia research and cytogenetics, which fully confirmed the prediction of Boveri in 1914 that cancer would prove to be a disorder involving chromosomes, and which also showed the relevance of cytogenetics, and genetics as a whole, to disorders of somatic cells, as well as to those constitutional conditions with which genetics had traditionally been associated.

References

1. Boveri T (1914). *Zur frage der Entstehung maligner tumoren*. Jena, Fischer (English translation, *On the problem of the origin of malignant tumours*).
2. Harris H (1995). *The Cells of the Body. A History of Somatic Cell Genetics*. Cold Spring Harbor, CSHL Press.
3. Levan A (1956). Chromosomes in cancer tissue. *Ann. N.Y. Acad. Sci.* **63**, 774–792.
4. Hauschka T (1963). Chromosome patterns in primary neoplasia. *Exper. Cell Res.* **24** (Suppl. 9), 86–98.
5. Levan A (1956). Chromosome studies of some human tumors and tissues of normal origin, grown in vivo and in vitro at the Sloan-Kettering Institute. *Cancer* **9**, 648–663.
6. Tjio JH and Levan A (1956). The chromosome number of man. *Hereditas* **42**, 1–6.
7. Andres AH and Shiwago PI (1933). Karyologische studien an myeloischer Leukämie des Menschen. *Folia haemat.* **49**, 1–20.
8. Chrustschoff GK, Andres AH and Ilina-Kakujewa WI (1931). Kulturen von blutleukozyten als methode zum stadium des menslichen karyotypus. *Anat. Anz.* **73**, 159–168.
9. Ford CE, Jacobs PA and Lajtha LG (1958). Human somatic chromosomes. *Nature* **181**, 1565–1568.
10. Baikie AG, Court Brown WM, Jacobs PA and Milne JS (1959). Chromosome studies in human leukaemia. *Lancet* **2**, 425–428.
11. Nowell PC (1993). From chromosomes to oncogenes: a personal perspective. In: Kirsch IR (ed.), *The Causes and Consequences of Chromosomal Aberrations*. Boca Raton, CRC Press, pp. 505–516.
12. Nowell PC (2002). Progress with chronic myelogenous leukaemia: a personal perspective over four decades. *Ann. Rev. Med.* **53**, 1–13.
13. Nowell (1960). Phytohemagglutinin: an initiator of mitosis in cultures of normal human leucocytes. *Cancer Res.* **20**, 462–467.
14. Nowell PC and Hungerford DA (1960). Chromosome studies on normal and leukemic human leukocytes. *J. Nat. Cancer Inst.* **25**, 85–93.
15. Nowell PC and Hungerford DA (1960). A minute chromosome in human chronic granulocytic leukaemia. (Abstract). *Science* **132**, 1497.
16. Nowell PC and Hungerford DA (1961). Chromosome studies in human leukaemia II. Chronic granulocytic leukaemia. *J. Nat. Cancer Inst.* **27**, 1013–1035.
17. Baikie AG, Court Brown WM, Buckton KE, Harnden DG, Jacobs PA and Tough IM (1960). A possible specific chromosome abnormality in human chronic myeloid leukaemia. *Nature* **188**, 1165–1166.
18. Tough IM, Court Brown WM, Baikie AG, Buckton KE, Harnden DG, Jacobs PA, King MJ and McBride JA (1961). Cytogenetic studies in chronic myeloid leukaemia and acute leukaemia associated with mongolism. *Lancet* **1**, 411–417.
19. Harnden DG (1996). Early studies on human chromosomes. *BioEssays* **18**, 163–168.
20. Rowley JD (1973). A new consistent chromosomal abnormality in chronic myelogenous leukaemia identified by quinacrine fluorescence and Giemsa staining. *Nature* **243**, 290–293.
21. Nowell PC, Rowley J and Knudson A (1998). Cancer genetics, cytogenetics – defining the enemy within. *Nature Med.* **4**, 1107–1111.
22. Hungerford DA (1978). Some early studies of human chromosomes, 1879–1955. *Cytogenet. Cell Genet.* **20**, 1–11.

Addendum

Nowell PC and Hungerford DA (1960). Chromosome studies on normal and leukemic human leukocytes. *J. Nat. Cancer Inst.* **25**, 85–93.
Reproduced with permission from Oxford University Press.

The 16 pages of plates accompanying this article have been omitted, apart from that demonstrating the abnormal minute chromosome, shown in the text as Figure 6-3.

Chromosome Studies on Normal and Leukemic Human Leukocytes [1]

PETER C. NOWELL,[2] *and* DAVID A. HUNGERFORD,[3,4]
Department of Pathology, School of Medicine, University of Pennsylvania, and The Institute for Cancer Research, Philadelphia, Pennsylvania

SUMMARY

Chromosome studies were made on myeloblasts obtained from the peripheral blood of 4 patients (3 males, 1 female) with acute or chronic granulocytic leukemia and on leukocytes from 3 healthy individuals (1 male, 2 females). The cells were grown in culture for 2 to 4 days before examination. Both in normal and in leukemic cells, the predominant metaphase chromosome number was 46. However, idiogram analyses revealed a definite abnormality, probably involving the Y chromosome, in the 2 cases of chronic granulocytic leukemia. No chromosome abnormality was distinguishable in the 2 cases of acute leukemia (1 male, 1 female) investigated. The results indicate that in some leukemic human leukocytes small but definite chromosomal changes are demonstrable.—J. Nat. Cancer Inst. 25: 85–109, 1960.

ALTHOUGH THERE have been numerous reports in recent years of chromosome abnormalities in a wide variety of malignant tumors, there is very little information on the chromosomes of leukemic cells in man. A few workers (*1, 2*) have observed mitotic figures in direct smears or in tissue cultures of cells from the blood or bone marrow of patients with leukemia and have noted that in general the chromosome number appeared to be in the diploid range; earlier, Andres and Shiwago (*3*) described mitotic variations in leukemic cells similar to those seen in other malignant cells. However, until three very recent reports, no exact counts or analyses had been made.

Nowell *et al.* (*4*) obtained myeloblasts from the peripheral blood of 3 patients with acute granulocytic leukemia and, after short-term culture, found that the chromosome number agreed with the diploid value for man ($2n = 46$) and that no gross chromosomal abnormalities were present. More recently, Ford *et al.* (*5*) reported similar results from studies of bone marrow in short-term culture in one case of lymphocytic leukemia and

---

[1] Received for publication December 14, 1959.
[2] Supported in part by Senior Research Fellowship SF-4 from the Public Health Service and research grant C-3562 from the National Cancer Institute, National Institutes of Health, Public Health Service.
[3] Supported by a research grant from the American Cancer Society, Inc., to Dr. Jack Schultz.
[4] The authors wish to express their appreciation to Miss Juliet Goodfriend, Miss Sarah Reifsnyder, and Mrs. Elizabeth Krohnert for technical assistance.

in another of blast-cell leukemia, but, in a third leukemia, of unspecified type, an abnormal number of 44 chromosomes plus 1 minute was observed. Baikie *et al.* (*6*) have just described chromosome studies in 11 cases of leukemia, employing the bone marrow technique. No aberrations were found in preliminary investigation of 6 cases of chronic leukemia, but in 5 cases of acute leukemia, there was abnormality of chromosome number in 1 case, abnormality of chromosome morphology in 2 cases, and abnormality of both number and morphology in the 1 case reported in detail.

In the present study, which is an extension of our previous report, the chromosomal characteristics of both normal and leukemic human leukocytes have been investigated by the use of short-term primary cell cultures, and it is believed, for reasons which will be discussed, that the results obtained accurately reflect the state *in vivo*. One of the 3 normal individuals described here (Normal case 1) was included in our earlier abstract, as were the 2 acute leukemia cases (1 and 2). In these cases, the preliminary data have been confirmed and amplified. Two additional normal cases and and 2 chronic leukemia cases are described here for the first time.

MATERIALS AND METHODS

Leukocytes were separated with phytohemagglutinin (Difco) from heparinized peripheral blood of healthy individuals and of patients with granulocytic leukemia. The cells were grown in short-term culture by a modification of Osgood's "gradient" method, which has been reported in detail elsewhere (*7, 8*). Briefly, cells were grown in undisturbed deep cultures in a mixture of autogenous plasma and commercial medium (TC-199, Difco). Each culture bottle contained 1 or 2 slanted slides, and the cells settled out and grew on the slides as well as on the sides and bottom of the bottle. Mitoses appeared in the leukemia cultures on the 1st or 2d day, but were rarely observed in the cultures from normal individuals before the 3d day. In the leukemia cultures, nearly all of the inoculated cells were immature myeloid forms and were the source of the mitotic figures studied. The dividing cells in the normal cultures were derived from monocytes and perhaps large lymphocytes (*7, 8*). Recent work (*9*) indicates that the phytohemagglutinin used in separating the leukocytes from whole blood is a specific initiator of mitotic activity among the normal leukocytes in these cultures.

Colchicine (1×10^{-7}M) was added at the time the cells were undergoing their first division in culture. Seventeen to 19 hours later, the cells were harvested from the bottom of the culture bottle, pretreated, fixed and stained in acetic orcein, and squash preparations were made (*7, 10*). At the same time, the slides from the culture bottles were rinsed in distilled water, air-dried, and stained with Giemsa. The slides were used only for general survey purposes, while the actual chromosome counts and analyses were made on the squash preparations. In some instances, colchicine was not used, so that anaphases and details of uncontracted metaphase chromosomes could be examined.

Idiograms were constructed from enlarged photomicrographs. In each case, the chromosomes have been classified into 3 groups, according to centromere position: median (top row, 7 pairs), subterminal (middle row, 6 pairs), and submedian (bottom row, 10 pairs). Within each of the 3 groups, the pairs have been arranged according to approximate order of decreasing length.

CASE HISTORIES

Chromosome studies were made on the leukocytes of 4 patients with acute or chronic granulocytic leukemia and 3 healthy individuals, members of the laboratory staff. Brief histories of the leukemia cases follow:

Leukemia case 1.—E. C., a 59-year-old white male, was admitted to Philadelphia General Hospital on July 25, 1957, because of weakness and fever of 1 month's duration. WBC = 37,000 (64% blasts); bone marrow diagnosis: "acute granulocytic leukemia." He was treated with 6-mercaptopurine and steroids during July and August, 1957, and from November, 1957, until his death, but never attained complete remission. White counts fluctuated from 3,000 to 130,000, with a high percentage of blast forms at all times. He died on April 17, 1958, after recurrent episodes of hemorrhage and infection. A total of 3 blood samples was obtained for chromosome study from this patient: 2 midway in the course of his disease (November 22, 1957, and December 2, 1957) and 1 during the terminal stages (March 28, 1958). White blood cell counts at those times were, respectively, 30,000 (92% blasts); 35,000 (88% blasts); and 52,000 (84% blasts).

Leukemia case 2.—I. K., a 35-year-old white female, was admitted to the Hospital of the University of Pennsylvania on October 12, 1957, with a history of weakness and purpura for 3 weeks. WBC = 12,000 (85% blasts and promyelocytes); bone marrow diagnosis: "acute granulocytic leukemia." Treatment with 6-mercaptopurine and steroids was started but was without effect. She died on October 16, 1957, with multiple hemorrhages. A blood sample for chromosome study was obtained on October 14, 1957; at that time WBC = 11,000 (91% blasts and promyelocytes).

Leukemia case 3.—E. K., a 41-year-old white male, whose case was first diagnosed as "chronic granulocytic leukemia" in April, 1955 (WBC = 318,000), was treated with Myleran and a remission was obtained which lasted until February, 1957. Exacerbation of the leukemia at that time was treated, with only partial success, by X-ray therapy to the long bones and with P^{32}. He remained reasonably well until July, 1958, when a second exacerbation of the disease failed to respond to further X-ray and Myleran therapy. He was admitted to the Hospital of the University of Pennsylvania August 13, 1958, and bone marrow examination confirmed the diagnosis of chronic granulocytic leukemia. He died of cerebral hemorrhage on August 27, 1958. A blood sample for chromosome study was obtained on August 19, 1958; at that time WBC = 375,000 (blasts and promyelocytes, 16%; myelocytes, 37%; metamyelocytes, 33%; segmenters, 12%; lymphocytes, 2%; monocytes, 0%).

Leukemia case 4.—G. H., a 33-year-old Negro male, was admitted to Philadelphia General Hospital on August 13, 1958, with a history of weakness and fatigue for 2 months. WBC = 248,000; bone marrow diagnosis: "chronic granulocytic leukemia." Myleran therapy was started on August 22, 1958. He was discharged, in remission, on September 17, 1958, and lost to follow-up. Blood for chromosome study was obtained on August 25, 1958; at that time WBC = 222,000 (blasts and promyelocytes, 17%; myelocytes, 16%; metamyelocytes, 14%; bands and segmenters, 31%; eosinophils, 7%; basophils, 10%; lymphocytes, 5%; monocytes, 0%).

RESULTS

Preliminary study of Giemsa-stained slides from both leukemia and normal cultures, neither of which had received colchicine, failed to reveal any gross mitotic abnormalities. Anaphases were entirely regular, without lagging or bridging of chromosomes. There were no spindle abnormalities such as multipolarity. Metaphase chromosome numbers were predominantly in the diploid range. Approximate counts of 200 metaphases on one slide (Leukemia case 1) gave a mean value of 44 chromosomes, with a standard error of ± 4. In both the normal and the leukemia cultures, approximately 2 percent of the metaphases were in the tetraploid range.

More precise studies on the squash preparations confirmed the impressions gained from examination of the Giemsa-stained culture slides. Chromosome counts were done on at least 30 metaphases in each case (tables 1 and 2); only those counts of metaphases which were accurate within 1 chromosome were accepted. A count recorded in the tables as 45 ± 1, 46 ± 1, 47 ± 1, or 48 ± 1 indicates that at one place in the metaphase plate it could not be definitely decided whether 1 or 2 chromosomes were present. The predominant chromosome number was 46 in all cases, both normal and leukemic.

TABLE 1.—Metaphase chromosome numbers of leukocytes from peripheral blood of normal humans

Case number	Chromosome number					
	45	45 ± 1	46	46 ± 1	47 ± 1	47
1	3	3	25	4	4	—
2	—	3	32	2	1	3
3	—	2	26	1	3	—

TABLE 2.—Metaphase chromosome numbers of leukocytes from peripheral blood of leukemic humans

Case number	Chromosome number						
	44	45 ± 1	46	46 ± 1	47 ± 1	47	48 ± 1
1	—	11	21	5	5	3	—
2	—	10	13	5	6	—	—
3	1	9	11	2	7	—	—
4	—	8	16	7	9	1	1

The tetraploidy which had been observed in both the normal and leukemia cultures was found, in the squash preparations, to be of two types, one showing endoreduplication (*11*) and one not; none of these was accurately countable and they are not shown in the tables. In addition, occasional apparently intact metaphases were observed in colchicine-treated material, *both* normal and leukemic, which had markedly *hypo-*

CHROMOSOME STUDIES IN HUMAN LEUKEMIA 89

diploid chromosome numbers. Such a cell, with 28 chromosomes, is shown in figure 1. In one individual, not included in this report, idiogram analyses were made of 3 such metaphases, each having 10 chromosomes. There was little or no correspondence of chromosome types among the 3 and, in general, no pattern either of chromosome number or type was observed in other hypodiploid metaphases; their chromosome complements appeared to be derived at random from the normal complement. Since these hypodiploid forms were never observed in the absence of colchicine, it is thought that they represent a phenomenon induced by the relatively long colchicine treatments employed. Study of culture slides on which such cells appeared suggested that they resulted from a type of segmentation of cells arrested at metaphase. A stage in this process is shown in figure 2, a photomicrograph from a squash preparation; the 2 groups of chromosomes total 46.

These types of variation in chromosome number occurred with equal frequency in both normal and leukemic cultures. Of greater interest are the results of idiogram analysis, which revealed abnormalities of chromosome morphology apparently peculiar to cells in the 2 cases of chronic leukemia. Detailed descriptions of the results in each case follow.

Normal case 1.—White female, age 21. Metaphase chromosome numbers predominantly 46. Five idiogram analyses were made of cells with 46 chromosomes. Four of these were normal (figs. 3, 7), and 1 showed an abnormally long chromosome, apparently the result of a translocation. Of a total of 39 metaphases counted, 3 had 45 chromosomes. One of these was analyzed, and it was found that the missing chromosome belonged to the fifth or sixth longest pair of the group with submedian centromeres.

Normal case 2.—White female, age 48. Metaphase chromosome numbers predominantly 46. Three idiogram analyses were made of such cells and all were normal (figs. 4, 8). Of a total of 41 metaphases counted, 3 had 47 chromosomes. One of these was analyzed and found to have a large acentric or telocentric fragment in addition to a normal female complement (fig. 5).

Normal case 3.—Japanese male, age 31. Metaphase chromosome numbers exclusively 46, within the adopted limits of accuracy. A total of 31 metaphases was counted, of which 3 were analyzed (figs. 6, 9). No morphological irregularities were observed, and the Y chromosome (fig. 20a-c) compares well with descriptions of this chromosome by other workers (5, 13, 14).

Leukemia case 1.—White male, age 60. Metaphase chromosome numbers predominantly 46. Of 45 metaphases counted, 3 had 47 chromosomes. One of these 3 was analyzed, and an apparently normal complement was present, plus 1 minute chromosome. Five idiogram analyses were made of cells with 46 chromosomes and all were indistinguishable from normal (figs. 10, 15, 21a-c).

Leukemia case 2.—White female, age 35. Metaphase chromosome numbers exclusively 46, within the error tolerated. Thirty-four mettaphases were counted and 4 idiogram analyses were made; all were indistinguishable from normal (figs. 11, 16).

Leukemia case 3.—White male, age 41. Thirty metaphases were counted. Chromosome numbers were 46, within error tolerated, with the exception of a single count of 44 (idiogram analysis not possible). Analyses were made of 3 cells with 46 chromosomes. All idiograms revealed the absence of a normal Y and the presence, in its stead, of a chromosome less than half the size of the smallest autosomes (figs. 12, 17, 22a–c). This minute chromosome was recognized in many other metaphases that were not subjected to idiogram analysis.

Leukemia case 4.—Negro male, age 33. In 42 metaphases counted, chromosome numbers were predominantly 46, with 1 count of 48 ± 1 (idiogram not possible). Analyses were made of 5 cells with 46 chromosomes. One idiogram (figs. 14, 19) showed an apparently normal Y chromosome. The other 4 idiograms, however, showed, instead of the Y, a minute chromosome (figs. 13, 18, 23a–c) similar to that observed in Leukemia case 3.

Figures 20 to 23 provide a comparison of Normal case 3 (male) with Leukemia cases 1, 3, and 4 (males). The comparison is made to illustrate the relative sizes of the two smallest pairs of acrocentric chromosomes and the Y chromosome or the minute chromosome, whichever was present in each metaphase.

DISCUSSION

Although the metaphase chromosome numbers were predominantly 46 in all 4 cases of granulocytic leukemia investigated, definite abnormalities of chromosome morphology were present in 2 of the individuals. In the 2 cases of chronic leukemia in males, idiogram analyses revealed changes of a similar type involving the presence of a minute chromosome. These minutes, which were morphologically similar but not identical in the 2 cases, were apparently present in place of a normal Y, and are judged to have resulted from either a deletion of a portion of the Y chromosome or the replacement of the Y by an autosomal fragment. The possibility that the minute simply represents normal variation in the morphology of the Y chromosome seems unlikely since studies by other workers have not revealed such variation (*12, 13*). Nor does it seem likely that the minute resulted from an abnormality of 1 of the 4 smallest acrocentric autosomes rather than of the Y, since none of the chromosomes interpreted in our material as being these 4 autosomes resembled a normal Y, such as was observed in our Normal case 3 and has been described by others (*12, 13*). Interestingly, if the change observed does involve the Y chromosome, it is not correlated with sex linkage in the incidence of chronic granulocytic leukemia, since statistical studies show only slightly greater frequency in males than in females (*14*).

The chromosomes of the 2 cases of acute leukemia (1 male, 1 female) were indistinguishable from normal with respect to both number and constitution. However, in the female, an alteration in chromosome morphology of the order of magnitude of that observed in the Y chromosome of the chronic leukemias might well be impossible to detect. In this connection it is, of course, possible that a variety of subtle cytogenetic changes [the "cryptostructural" rearrangements of Levan (15)] as well as purely genetic changes might be present in any of these leukemic cases and be undetectable by our methods.

On the basis of the few cases studied, no estimate can be made regarding the possible relationship of chromosome changes to the duration of the disease process or to the duration or type of therapy. However, in Leukemia case 1 no chromosome changes were present even after prolonged chemotherapy, while in Leukemia case 4 the minute chromosome was present although treatment had just been started. In the one case reported in detail by Baikie et al. (6), a female with acute granulocytic leukemia, the modal chromosome number changed after 7 months of steroid therapy, but abnormalities of both number and morphology were present before treatment was begun.

There seems little doubt that the "leukemic" chromosome patterns studied in the present paper were indeed derived from leukemic cells. Nearly all of the cells planted in the cultures from leukemic patients were immature myeloid forms; monocytes and large lymphocytes, the cells which divide in *normal* cultures, constituted no more than 5 percent of the original inoculum. Furthermore, the leukemic mitoses studied had usually occurred during the first 2 days in culture, whereas normal leukocytes were rarely observed in mitosis before the 3d day of culture. Finally, when permitted to mature *in vitro*, the dividing cells in the leukemia cultures developed into polymorphonuclear forms, while the dividing cells in the normal cultures differentiated into macrophages, multinucleated giant cells, and small lymphocytes (8).

The present study also provides data on the chromosomes of leukocytes of three normal individuals, one male and two female. Of particular interest in these cases is the low but definite incidence of chromosome abnormalities in these normal cells. Such abnormalities among other populations of normal human somatic cells have been reported previously by other authors. Counts other than euploid are regarded by Ford et al. (5) as artifacts of technique or interpretation. Tjio and Puck (12) state that of 2,000 cells, 99.9 percent had chromosome numbers of 46 or 92, with a tetraploid frequency of less than 3 percent. Chu and Giles (13) mention that in a few exceptional cells deviations of one chromosome from normal were observed. These were interpreted as representing either instances of somatic aneuploidy, originally present or induced during culture, or artifacts resulting from errors in technique. In the three normal cases presented here, there is no suspicion of pathological origin of the cells studied. The chance that such aneuploid numbers were induced during culture is remote, since only the first or, at the very most, the second mitosis *in vitro* is represented in these preparations. We believe that technical and interpretive errors have been ruled out in the cases of aneuploidy and other chromosome aberrations in which idiogram analyses were made. While, as mentioned, it is not possible to estimate reliably in our preparations the frequency of such aberrations, it is reasonable to believe it is greater than 0.1 percent and probably not more than 2 percent.

The relationship of the present findings in human leukemia to the results of such investigations on other types of malignant tumors is not yet clear. Most previous studies of spontaneous mammalian cancers have revealed much greater variability of chromosome number and morphology than that reported here [see, for example Hsu (16), Koller (17)]. Leukemic cells are apparently not so far removed from their normal prototypes as the cells of many types of malignant tumors. For instance, leukemic "blasts" are essentially indistinguishable, both morphologically and biochemically (18), from the "blasts" of normal bone marrow, and *normal* as well as leukemic leukocytes have the capability, present only in *malignant* cells of other tissues, of detaching from their fellows and spreading throughout the body. Whether the more subtle chromosome changes observed in leukemic cells, as compared to other types of malignancies, are related to these biological differences has not yet been determined. Furthermore, the relationship of our results to those of Baikie et al. (6), who observed chromosomal abnormalities in acute leukemia, but not in chronic leukemia, cannot as yet be evaluated. The studies of chronic leukemia reported by Baikie et al. (6) were of a preliminary nature and might not have revealed the small chromosome abnormality observed in our chronic cases. Obviously, more extensive studies are needed on the chromosomes of various types of human leukemia and lymphoma, and such studies are planned. In addition, cytological investigations of other early primary neoplasms, both benign and malignant, such as the recent study by Palmer (19) of the Shope rabbit papilloma, may lead to a better understanding of the relationship of chromosome changes to the pathogenesis of tumors.

Notes added in proof:

1) A recent paper by Bayreuther (Nature 186: 6–9, 1960) indicates that the incidence of chromosome aberrations in a variety of early spontaneous and induced malignancies may not be as great as indicated by other workers.

2) A more detailed analysis of chromosomes in human leukemia will be provided by studies now in progress in our laboratories, employing an improved method of chromosome preparation from leukocyte cultures (Moorhead, P. S., Nowell, P. C., Mellman, W. J., Batipps, D. M., and Hungerford, D. A.: To be published).

REFERENCES

(1) BARIGOZZI, C.: Les chromosomes humains dans quelques états pathologiques. Arch. Julius Klaus Stift. 22: 342–345, 1947.

(2) GUNZ, F. W.: Culture of human leukaemic blood cells *in vitro*; normal and abnormal cell division and maturation. Brit. J. Cancer 2: 41–48, 1948.

(3) ANDRES, A. H., and SHIWAGO, P. I.: Karyologische Studien an myeloischer Leukämie des Menschen. Folia haemat. 49: 1–20, 1933.

(4) NOWELL, P. C., HUNGERFORD, D. A., and BROOKS, C. D.: Chromosomal characteristics of normal and leukemic human leukocytes after short-term tissue culture. (Abstract.) Proc. Am. Assoc. Cancer Res. 2: 331–332, 1958.
(5) FORD, C. E., JACOBS, P. A., and LAJTHA, L. G.: Human somatic chromosomes. Nature, London 181: 1565–1568, 1958.
(6) BAIKIE, A. G., COURT BROWN, W. M., JACOBS, P. A., and MILNE, J. S.: Chromosome studies in human leukaemia. Lancet ii: 425–428, 1959.
(7) HUNGERFORD, D. A., DONNELLY, A. J., NOWELL, P. C., and BECK, S.: The chromosome constitution of a human phenotypic intersex. Am. J. Human Genet. 11: 215–236, 1959.
(8) NOWELL, P. C.: Differentiation of human leukemic leukocytes in tissue culture. Exper. Cell Res. 19: 267–277, 1960.
(9) ———: Phytohemagglutinin: An initiator of mitosis in cultures of normal human leukocytes. Cancer Res. 20: 462–466, 1960.
(10) HUNGERFORD, D. A., and DIBERARDINO, M.: Cytological effects of prefixation treatment. J. Biophys. & Biochem. Cytol. 4: 391–400, 1958.
(11) LEVAN, A., and HAUSCHKA, T. S.: Endomitotic reduplication mechanisms in ascites tumors of the mouse. J. Nat. Cancer Inst. 14: 1–43, 1953.
(12) TJIO, J. H., and PUCK, T. T.: The somatic chromosomes of man. Proc. Nat. Acad. Sc. 44: 1229–1237, 1958.
(13) CHU, E. H. Y., and GILES, N. H.: Human chromosome complements in normal somatic cells in culture. Am. J. Human Genet. 11: 63–79, 1959.
(14) GILLIAM, A. G.: Age, sex, and race selection at death from leukemia and the lymphomas. Blood 8: 693–702, 1953.
(15) LEVAN, A.: Chromosomes in cancer tissue. Ann. New York Acad. Sc. 63: 774–792, 1956.
(16) HSU, T. C.: Numerical variation of chromosomes in higher animals. In Developmental Cytology (Rudnick, D., ed.). New York, Ronald, 1959, pp. 45–62.
(17) KOLLER, P. C.: Cytological variability in human carcinomatosis. Ann. New York Acad. Sc. 63: 793–817, 1956.
(18) LAJTHA, L. G.: Bone marrow cell metabolism. Physiol. Rev. 37: 50–65, 1957.
(19) PALMER, C. G.: The cytology of rabbit papillomas and derived carcinomas. J. Nat. Cancer Inst. 23: 241–250, 1959.

CHAPTER 7

Technology and nomenclature: the next steps, 1960

THE SUDDEN EXPLOSION OF KNOWLEDGE about human chromosomes that had hit the world during 1959 and 1960 made it clear to the wider medical and scientific community that human cytogenetics was becoming a major clinical as well as research discipline, with applications in paediatrics and adult disorders, and also in leukaemia investigations. Yet at this point the ability to analyse human chromosomes and to culture suitable tissue for this was limited to a handful of laboratories worldwide, all orientated to research, and mostly using material, such as bone marrow or testis, which was too invasive to obtain to form the basis for systematic laboratory diagnosis.

This chapter describes the main initial developments that were necessary to bring human cytogenetics into clinical laboratory medicine; they occurred very rapidly and were equally quickly disseminated worldwide, so that by the end of 1960 numerous laboratories were able to begin providing diagnostic cytogenetic analysis using techniques that would remain largely unchanged for the next decade.

Workers in the field also faced another problem, seen when any new discipline suddenly expands; the explosion of information produced problems of nomenclature and classification, threatening to impede clear communication between groups and to transmit confusion rather than understanding to the wider world. 1960 was also the year when these issues were confronted and, to a considerable extent resolved, a process which proved to be as important to human cytogenetics as the new technology.

The simplification of human chromosome analysis

Skin fibroblast culture

Earlier chapters have described how cultured bone marrow was already in use for chromosome studies before 1960, but this invasive technique was not suitable for young children, nor as a regular diagnostic investigation. The use of cultured fibroblasts for cytogenetic analysis was also not entirely new; Tjio and Levan[1] had used cultured embryonic lung fibroblasts,

Fig. 7-1 A 'cytogenetic self-portrait' based on cultured skin fibroblasts. (Courtesy of Professor David Harnden.)

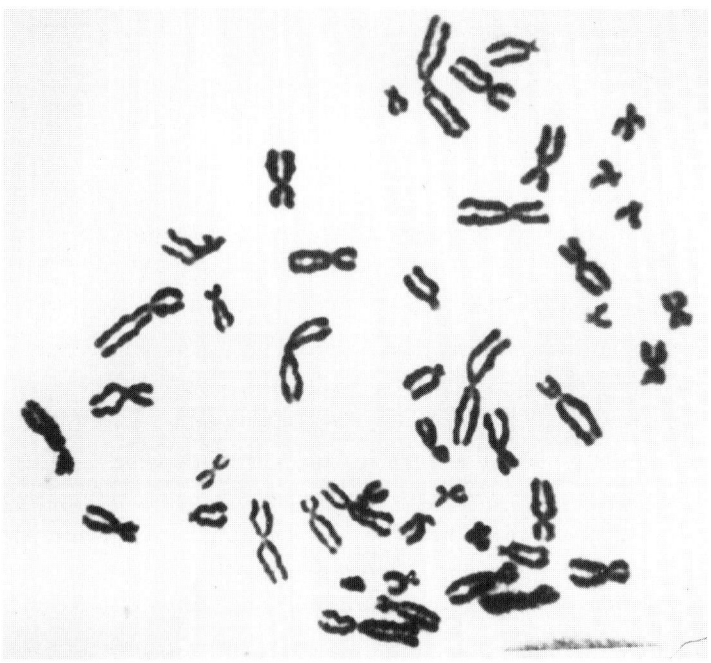

while Lejeune and colleagues took biopsies from fascia lata;[2] but Harnden's development, published in 1960, of a simple culture method that could be used on a small skin biopsy[3] was a major step in widening access to cytogenetic analysis. His 'cytogenetic self-portrait' (Figure 7-1) is certainly the first such documented karyotype, though others (including Marthe Gautier) were by then also using skin biopsies from themselves as controls. Harnden's method was also helped by the clinical technique of John Edwards, who found that a superficial 'pinch' of skin taken with forceps could replace the previously used full thickness 'punch biopsy'.

Although the need for skin biopsy in most situations was soon removed by peripheral blood culture methods (see below), it still remains important in the detection of mosaicism.

General improvements in tissue and cell culture techniques were also important; although these had started to find application in human cytogenetics during the 1950s, as noted in Chapter 1, it was the simplification and standardisation of these by workers such as Theodore Puck and Ernest Chu that proved especially fruitful. Both have given historical accounts of how their work developed;[4,5] Puck's role was also important in providing Joe Hin Tjio with an initial base in America and the quality of chromosome analysis resulting from their joint work can be seen in their joint publications,[6,7] which greatly extended his original observations with Levan.

Peripheral blood for chromosome analysis

Of all the technological advances described in this chapter and in Chapter 1, the ability to use a simple venous blood sample has been the one that has done most to promote clinical cytogenetics, allowing clinicians with no knowledge of chromosomes and no laboratory experience to take and send samples from their patients with minimal discomfort, often as part of wider medical investigations. It is hardly surprising that blood should have rapidly replaced the use of cultured skin fibroblasts; what is surprising is that it took until 1960 for this to happen and that the initial work on human chromosome abnormalities still had to be dependent on more invasive procedures when, almost 30 years before, the main elements of methods for the chromosome analysis of human blood had already been fully described.

JBS Haldane had recognised the problem in 1932, in his paper predicting the existence of XXY females on the basis of general genetic analysis,[8] concluding that:

....a technique for the counting of human chromosomes without involving the death of the person concerned is greatly to be desired. It seems possible that satisfactory mitoses might be observed in a culture of leucocytes. If so, the development of human cytology in relation to genetics will become possible.

In fact, just such a technique was already being developed in Moscow, a preliminary report (in German) having appeared in 1931,[9] leading to a full, detailed and clear account by Chrustschoff and Berlin in 1935.[10] The paper not only described the culture methods, including use of embryo extract, but used hypotonic solutions (again almost 30 years before TC Hsu's work, as noted in Chapter 1) and it also recognised the importance of autolysing the red blood cells, suggesting that this released substances which stimulated mitosis in the white cells. The paper could not be clearer on the relevance of this last aspect:

In analysing the problem of the factors stimulating the mutation of non-granular forms of blood into stable cells of united tissue type, we must call attention to an extremely important point. In our first communication we pointed out that the number of erythrocytes in the cultivated fragment of the blood clot exercises an influence on the process of transformation. We have repeatedly observed the same thing in all the experiments made after the publication of our first communication. On the ground of these observations we advance the hypothesis that some products of the disintegration of blood cells, and especially of erythrocytes, are an essential factor stimulating and possibly determining the process of transformation.

In all cases we obtained a definite and uninterrupted effect accelerating the mutation of the non-granular forms that have migrated into the medium.

It has to be said that the chromosome counts resulting from the work were variable and inaccurate (around 52 chromosomes), though this was not the main focus of the work.

It seems remarkable that such an important paper, fully illustrated and appearing in clearly written English in a high-profile international genetics journal (*Journal of Genetics*), edited by Haldane himself, should have been so completely ignored by most later workers, especially since during the 1930s Russia was recognised as being in the forefront of genetics research. Nor was it an isolated contribution, but rather a systematic programme, involving a large team, that covered the key areas of cytogenetic research at the time, as outlined in Chapter 1.

However sad the ignoring of the Russian work in the West and the fate of the workers involved may have been, it was inevitable that the use of peripheral blood would again become the focus of attention, and not surprising that a haematologist should have been responsible. It was the leukaemia work of Nowell and his colleagues in Philadelphia that finally made the systematic use of peripheral blood feasible.[11] In addition to David Hungerford, Paul Moorhead, who had previously worked with TC Hsu, had come to Philadelphia, providing further cell culture expertise.

Nowell, in a retrospective review of his own career,[12] freely admits that he was unaware of this earlier work on leucocyte culture and cytogenetics. His own group's approach was based on the somewhat different 'gradient' culture method described by Osgood,[13] but also included a key step that was novel in its application, even though already known for other uses. This was the use of the substance phytohemagglutinin, which was to become a key part of modern cytogenetic technique.

Phytohemagglutinin (PHA)

Phytohemagglutinin was the name given to a mucoprotein found in bean extracts that was characterised by Osgood and his colleagues[14] and used by them in their gradient culture methods to remove the red blood cells. It was already recognised by the Russian workers that haemolysis of the red cells was important if the white cells were to be stimulated to divide, but whereas they had considered that some stimulatory substance was released by the process of haemolysis, phytohemagglutinin acted, as its name implies, by agglutination.

Nowell's particular contribution[15] was to recognise that agglutination was not the only effect of phytohemagglutinin. At this point, most of the samples he was studying were from leukaemic patients with dividing white cells in their peripheral blood. Unaware of the earlier Russian work, he was not expecting mitoses in normal blood. In his 'personal perspective' he recounts:

One day my technician and I travelled to the Presbyterian Hospital, where I had trained briefly as a resident, to obtain some leukemic blood. We found that the patient was in remission, but rather than waste the trip we cultured the peripheral blood leukocytes anyway. To our surprise, we found many mitotic figures. This led to the culturing of my own cells, with similar results, and the conclusion that either I had leukemia or we had found a way to get normal peripheral blood leukocytes to divide in culture.

...after eliminating one by one the variables in the culture system, it eventually became clear that the mitogen was the PHA we had been using to remove the erythrocytes. I was disappointed that the stimulant was not more physiologic, and tended to agree with one reviewer of the resultant manuscript submitted to Cancer Research, *that it was an interesting observation but of no obvious importance.*

Happily, the second alternative explanation for the mitoses proved to be correct, while phytohemagglutinin, contrary to the expectation of the reviewer and of Nowell himself, rapidly became a standard element in peripheral blood cytogenetics, an indication of how difficult it is to predict in advance the applicability of many new developments.

It is also worth noting that at this time no one had any knowledge of how phytohemagglutinin acted in stimulating mitosis, nor even which type of white blood cell was involved. Nowell's conclusion that this must be the large lymphocyte or monocyte, based on 'the only logical experiment' (by his own estimate) in all of his early studies,[15] proved to be wrong when it was later shown that small lymphocytes could indeed undergo mitosis.[16] The different responses to phytohemagglutinin and to other mitogens provided key evidence in distinguishing T and B lymphocytes, thus giving theoretical as well as practical importance to what had originally been an entirely serendipitous innovation.

Air drying

This simple procedure deserves a note as yet another of the steps that helped to convert human cytogenetic analysis from a complex research activity into a routine diagnostic procedure. Again, it was not new, having already been described by Rothfels and Siminowitz,[17] but the Philadelphia group were able to incorporate it and eventually use it as a substitute for 'squashing' in their peripheral blood cultures,[11] achieving essentially the same result in terms of spreading chromosomes in a two-dimensional plane.

Nomenclature and classification

Some people regard naming and identification as tedious, albeit necessary tasks, but an agreed system of classification and nomenclature is essential in any field of science if it is to progress without confusion. The importance and permanent influence of such systems is in fact immense, as seen with Linnaeus' binomial system and with the naming of geographical landmarks by the early explorers. In genetics itself the whole process of human gene mapping has been underpinned by a systematic cartographic approach and by the continuous work of nomenclature committees, efforts which have rested to a considerable extent on the foundations of human cytogenetic nomenclature devised in 1960, as well as on catalogues of human inherited disorders, such as McKusick's *Mendelian Inheritance in Man*.

Already by this time there were the beginnings of multiple systems (at least six) for naming and classifying human chromosomes, mainly based on size and shape. That of the Paris workers,[18] based on a series of letters, is shown in Figure 7-2. Joe Hin Tjio had now, as mentioned, joined Theodore Puck in Denver, where the use of Puck's monolayer culture techniques had resulted in extremely high quality preparations, allowing clear distinction of many specific chromosomes. Tjio and Puck, like Ford, adopted a numerical system based on measurement, placing chromosomes in groups of comparable size and centromere position. Difficulties in separating autosomes from the sex chromosomes added to the problem, some workers including X and Y in their numbering system, while others kept them separate.

The 1960 Denver conference on nomenclature[19] proved to be a timely and, in the long run, extremely successful solution to these rapidly growing problems and the cytogenetics community certainly owes the organisers a very considerable debt for providing solid and satisfactory foundations that avoided many major later difficulties, as well as increasing the sense of community among those involved. In case any future science historian wishes to compare the success or failure of such initiatives in different fields, it is worth noting the main features of the meeting.

First, it was small – only 14 members and three moderators; those invited were almost all those who had directly observed and published a human karyotype. With hindsight, the only major worker omitted was Klaus Patau, from Madison, most of whose career had been in basic genetics and whose first published venture into human cytogenetics, as described in Chapter 5, was not published until later in 1960, after the conference. The meeting was international (seven members from Europe, six from America, (three of whom were of Chinese origin), one from Japan). Finally, there were three 'wise men' acting

Fig. 7-2 Nomenclature before the Denver Conference. Example (XXY karyotype) of the Paris system of Lejeune and colleagues. (Courtesy of Dr Marguerite Prieur.)

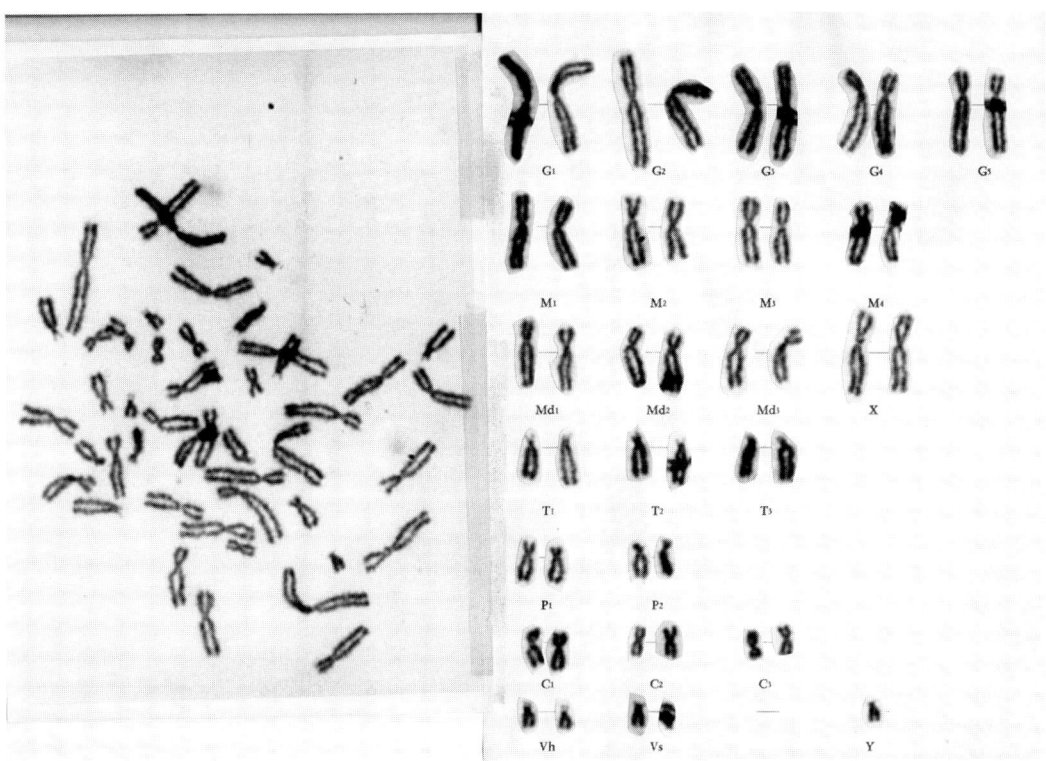

as moderators (Figure 7-3), all distinguished and respected geneticists, but with no involvement in cytogenetics; these were David Catcheside (Birmingham UK), Herman Muller (Indiana) and Curt Stern (Berkeley; Figure 7-4). The full list of participants can be seen in the report itself, which is reproduced at the end of this chapter.

The holding of the conference was suggested initially by Charles Ford, but its location in Denver, Colorado, under the auspices of Theodore Puck, pioneer of human cell culture, requires some explanation. After leaving Lund and Spain, Joe Hin Tjio had moved to Denver to work with Puck, with whom he had already collaborated, and also to obtain a PhD, necessary for an academic career in America. As noted above, their work had shown the need for a standardised nomenclature, but Tjio himself was not directly involved in the planning of the meeting, since he had moved again to Washington

Fig. 7-3 The Denver Conference, 1960. The 'three wise men'; conference moderators David Catcheside (centre), Herman Muller and Curt Stern (on Catcheside's left). (Photograph by JH Tjio, courtesy of Professor David Harnden.)

by the time it was held. Denver thus provided a relatively neutral venue, not being the base of any of the main protagonists in human cytogenetics.

Puck himself (Figure 7-5) has provided a vivid account of the meeting,[4] as have TC Hsu[20] and several others who were involved,[15,21] including three participants who I was able to interview (Marco Fraccaro, David Harnden and Patricia Jacobs). Puck's careful planning and preparation were almost sabotaged at the last minute by bureaucratic refusal of travel funding by the National Institutes for Health, but the American Cancer Society came to the rescue. Anyone who has organised a meeting will feel for Puck's dilemma.

I was sick with apprehension. The scientists were already on their way to Denver and I had no funds whatsoever to cover their travel expenses. I made a frenzied phone call to Dr Harry Weaver, vice president in charge of research of the American Cancer Society.... After about a minute of my desperate appeal, Dr Weaver cut me short by saying, 'Stop, I understand. You are now covered.'

Dr Weaver undoubtedly saved my sanity. He also won for the American Cancer Society the permanent distinction of having understood and supported the need for fundamental genetics in the coming era of medicine.[4]

Fig. 7-4 The Denver Conference, 1960; Patricia Jacobs with Curt Stern. (Photograph by JH Tjio, courtesy of Professor David Harnden.)

Fig. 7-5 Theodore T Puck, convener of the Denver Conference and pioneer of cell culture techniques in human cytogenetics. (Courtesy of Dr TT Puck.)

The role of the three eminent and respected moderators proved invaluable, but even so, it took three days of intense (apparently at times emotional) argument before agreement could be reached. There were essentially six different data sets (and naming systems) to be compared. According to David Harnden, the turning point was when, at some apparently insoluble impasse, Ernest Chu proposed a Chinese system for classification, which caused the others to abandon their differences. Thus, in the event a unanimous report was able to be produced and the moderators did not have to use their veto.

Both David Harnden in his historical paper[21] and TC Hsu in his book[20] (quoted below) give a little of the atmosphere of this remarkable meeting – and hint at some of the tensions that almost never appear in 'final communiqués', whether political or scientific.

The meeting lasted four days. Progress was made despite many heated disagreements. It was amazing to witness the emotional involvement over minute details.... The participants worked hard to reach a sensible, yet flexible, nomenclature system, and proceeded to write a report. The draft was read and reread, corrected and recorrected, and, by the afternoon of the third day, it was complete. The participants felt a sense of relief, as if a historic document was being written. Indeed it was; but thinking in retrospect, I could easily appreciate the difficulty of the American forefathers in arriving at the nation's constitution. Here we were worrying about how to name the 23 pairs of human chromosomes, not the welfare of the country and its people, yet it took three days to reach some agreement.

Looking at the report today (it was published in multiple journals,[19,22] which sometimes confuses citation), one's impressions are firstly how little of the basic classification has changed over the subsequent 45 years, but also how obvious and seemingly uncontroversial it is. This simplicity was the key to its universal acceptance and success, for its basis was the agreed measurements of the individual chromosomes in terms of total length, and the length and ratio of the chromosome arms. Terminal 'satellites' were excluded from the measurements and the sex chromosomes were left outside the numbering system. Only one chromosome might today be considered as 'misclassified'; the careful measurements of Levan and of Ferguson-Smith[23] showed that the chromosome trisomic in Down's syndrome was in fact the smallest and thus should have been ranked as 22, rather than as 21. By the time this was generally agreed (after the Denver conference), the general recognition of 'trisomy 21' made any change impossible.

Malcolm Ferguson-Smith, who at the time was developing a cytogenetics laboratory with Victor McKusick in Baltimore, refers to the problem in his historical review[23] and also remembers it in discussion in 2004:

I went down to Houston in 1959 and met Albert Levan and TC Hsu.... I showed Albert Levan my preparations and he showed me how to use the camera lucida to make pictures of chromosomes.

I persuaded him to come up to Hopkins some time in the Summer of 1959 and he came and brought his camera lucida equipment; we sat down in my 'cupboard' and we drew pictures of Down's syndrome chromosomes. He and I figured out that the extra one was the smallest and whereas Lejeune had called this chromosome '21' we said, this is a mistake – it is chromosome 22, you see. So I got into fearful trouble ever afterwards trying to convince everybody that it was chromosome 22 and not 21 – and of course it is!

Fig. 7-6 Nomenclature after the Denver Conference. Example (46XX karyotype) of the agreed system, with the additional A–G groupings suggested by Patau. (From Patau, 1961,[26] reproduced from *The Lancet* with permission from Elsevier.)

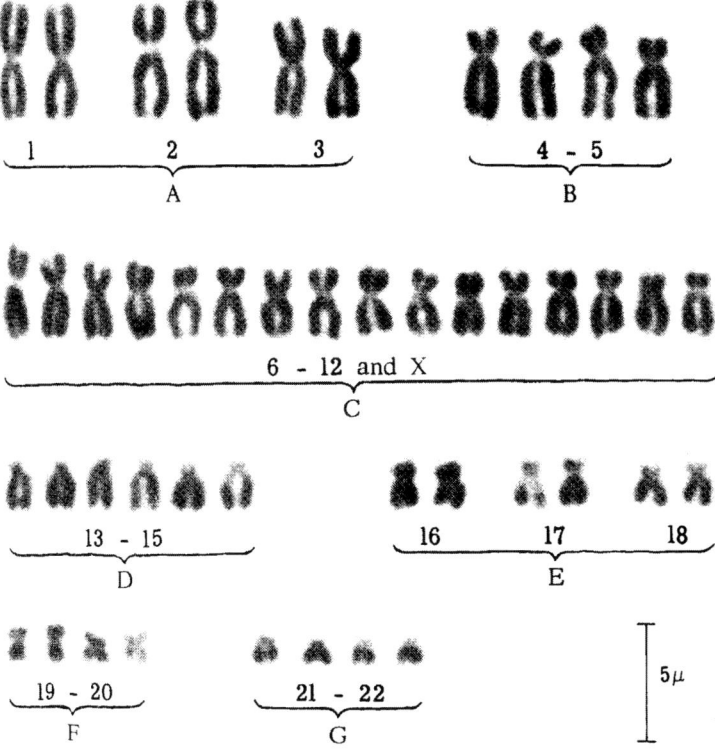

The only part of the Denver nomenclature that proved cumbersome was that for the chromosome groups (it has to be remembered that only the largest chromosomes could confidently be identified individually). Here, the subsequent suggestion of Patau in a rigorous and detailed paper[24] to use letters (A to G) rather than combined number (e.g. 6–12) to designate groups (Figure 7-6) proved more satisfactory and was adopted at the 1966 international nomenclature meeting in Chicago,[25] though again not without some difficulty according to John Hamerton, who chaired the meeting. Patau's incisive criticism and combative approach to the Denver system[26] must have made some members wish that they had had him 'on board' from the start, but one has a feeling that this might have resulted in the precariously agreed system breaking down completely! Patau's principal criticism, that many chromosomes given numbers could not be reliably distinguished, was undoubtedly a valid one; fortunately, once banding techniques allowed unequivocal identification of all human chromosomes individually, this aspect of nomenclature was no longer contentious.

The success of the Denver conference can be judged from the fact that it led to the establishment of the International System for Human Cytogenetic Nomenclature (ISCN) committee, created in 1978, which continues to meet regularly, and which has successfully incorporated later developments such as chromosome banding (dealt with in the ISCN report from the 1971 Paris meeting[27]), and *in situ* hybridisation, tackled in 1995.[28] Tharapel has given an account of how ISCN evolved from the original nomenclature meetings and of its development up to 1995.[29] Credit should also be given to the National Foundation March of Dimes for funding the early meetings, ensuring that organisers did not have to suffer the agonising of Puck before the Denver conference, and to Harold Klinger and Karger publishers, who provided a permanent published record, continued up to the present, in the journal *Cytogenetics* (later *Cytogenetics and Cell Genetics*). The most recent version appeared in 2005.[30]

One point, not dealt with at Denver, that readers may question, is the naming of the chromosome arms as p and q. It is generally agreed that Lejeune's influence led to the designation of the short arm as p (petit), but opinions differ as to the origin of q for the long arm. Harnden thinks it may have resulted from a mistyping of q for g (grand), but an alternative (possibly from more mathematically inclined geneticists) is that it simply represented the rest of the chromosome, on the basis that p + q = 1.

The topic of standardisation in human cytogenetics, including the Denver and succeeding conferences, has very recently been examined from the perspective of the social historian by Lindee,[31] who shows how the personal factors and structure of the professional communities involved interacted with the science itself. Her account illustrates what rich material the published and unpublished records represent, not only in this particular area, but across the whole range of human genetics, and emphasises the importance of preserving it.

In many ways 1960 marks a watershed, for it can be seen that by the end of this year, human cytogenetics had not only begun to make important medical contributions to congenital malformations and to somatic disorders such as leukaemia; it had also equipped itself with a series of techniques that simplified the original complex and often unpredictable methods, and that were adapted to routine use as part of medical laboratory investigations. The timely Denver conference had narrowly avoided an uncontrolled and confusing series of nomenclatures, and human cytogenetics could now begin a new life as a clinical discipline as well as a research tool.

The rapid proliferation of work, and of workers, after 1960, meant that some of the original intimacy shared by the small group of original pioneers would inevitably be lost, something that perhaps explains the strong links and friendships established at the Denver conference and maintained over the following decades. The conference, and the foundations that it helped to establish, rightly allowed this small group to feel a sense of collective pride in their efforts, in addition to that already resulting from their own work.

References

1. Tjio JH and Levan A (1956). The chromosome number of man. *Hereditas* 42, 1–6.
2. Lejeune J, Gautier M and Turpin R (1959). Les chromosomes humains en culture de tissus. *Comptes Rendues Acad. Sci.* 248, 602–603.
3. Harnden DG (1960). A human skin culture technique used for cytological examination. *Brit. J. Exper. Pathol.* 41, 31–37.
4. Puck TT (1994). Living history biography. *Am. J. Med. Genet.* 53, 274–284.
5. Chu EHY (2004). Early days of mammalian somatic cell genetics: the beginnings of experimental mutagenesis. *Mutation Res.* 566, 1–8.
6. Tjio JH and Puck TT (1958). The somatic chromosomes of man. *Proc. Nat. Acad. Sci. USA* 44, 1229–1237.
7. Tjio JH and Puck TT (1958). Genetics of somatic mammalian cells. II. Chromosomal constitution of cells in tissue culture. *J. Exper. Med.* 108, 259–268.
8. Haldane JBS (1932). Genetical evidence for a cytological abnormality in man. *J. Genet.* 26, 341–344.
9. Chrustschoff GK, Andres AH and Ilina-Kakujewa WI (1931). Kulturen von blutleukozyten als methode zum stadium des menslichen karyotypus. *Anat. Anz.* 73, 159–168.
10. Chrustschoff GK and Berlin EA (1935). Cytological investigations on cultures of normal human blood. *J. Genet.* 31, 243–261.
11. Moorhead P, Nowell P, Mellman W, Battips D and Hungerford D (1960). Chromosome preparations of leukocytes cultured from human peripheral blood. *Exp. Cell. Res.* 20, 613–616.
12. Nowell PC (1993). From chromosomes to oncogenes: a personal perspective. In: Kirsch IR (ed.) *The Causes and Consequences of Chromosome Aberration.* London, CRC Press, pp. 505–516.
13. Osgood EE and Krippachane ML (1955). The gradient tissue culture method. *Exp. Cell. Res.* 13, 1019.
14. Rigas DA and Osgood EE (1955) Purification and properties of the phytohemagglutinin of Phaseolus vulgaris. *J. Biol. Chem.* 212, 607–9.
15. Nowell P, (1960). Phytohemagglutinin: an initiator of mitosis in cultures of normal human leukocytes. *Cancer Res.* 20, 462–468.
16. Gowans JL, Gesner BL and McGregor DD (1961). The immunological activity of lymphocytes. In: Wolstenholme GEW and O'Connor M (eds), *Biological Activity of the Leukocyte.* London, CIBA Foundation, pp. 32–40.
17. Rothfels KH and Siminovitch L (1958). An air-drying technique for flattening chromosomes in mammalian cells grown in vitro. *Stain Technol.* 33, 73.
18. Lejeune J, Turpin R and Gautier M (1959). Le mongolisme, maladie chromosomique. *Bull. Acad. Nat. Méd.* 143, 256–265.
19. Denver Conference (1960). A proposed standard system of nomenclature of human mitotic chromosomes. *Lancet* 1, 1063–1065.
20. Hsu TC (1979). *Human and Mammalian Cytogenetics. An Historical Perspective.* New York, Springer.
21. Harnden DG (1996). Early studies on human chromosomes. *BioEssays* 18, 163–168.
22. Denver Conference (1960). *J. Hered.* 51, 214–221.
23. Ferguson-Smith MA (1993). From chromosome number to chromosome map: the contribution of human cytogenetics to genome mapping. *Chromosomes Today* 11, 3–19.
24. Patau K (1960). The identification of individual chromosomes, especially in man. *Am. J. Hum. Genet.* 12, 250–276.
25. Chicago Conference (1966). Standardization in Human Cytogenetics. Birth Defects Original Articles Series, 2, 2. National Foundation, New York.
26. Patau K (1961). Chromosome identification and the Denver Report. *Lancet* 1, 933–934.
27. Paris Conference (1971). Standardization in Human Cytogenetics. Birth Defects Original Articles Series, vol. 8, no. 7. National Foundation, New York.

28. Mitelman F (ed.) (1995). *ISCN 1995. An International System for Human Cytogenetic Nomenclature*. Basel, Karger.
29. Tharapel AT (1998). Evolution of human cytogenetic nomenclature and highlights of ISCN 1995. ECA Newsletter 1. www.biologia.uniba.it/eca/newsletter/news.
30. International System for Human Cytogenetic Nomenclature (ISCN) 2005 (2005). Basel, Karger.
31. Lindee S (2005). *Moments of Truth in Genetic Medicine*. Baltimore, Johns Hopkins University Press, pp. 90–119.

Addendum

The Denver Conference report. Published in a variety of journals, it is reproduced here from *The Lancet*, with permission from Elsevier.

MAY 14, 1960   SPECIAL ARTICLES   THE LANCET   1063

Special Articles

A PROPOSED STANDARD SYSTEM OF NOMENCLATURE OF HUMAN MITOTIC CHROMOSOMES

The following statement has been drawn up by a study group, which met to formulate an agreed system of nomenclature for human chromosomes, with a view to eliminating the confusion which has arisen from the independent publication of a variety of systems by different workers.

THE rapid growth of knowledge of human chromosomes in several laboratories, following advances in technical methods, has given rise to several systems by which the chromosomes are named. This has led to confusion in the literature and so to the need for resolving the differences. Consequently, at the suggestion of Dr. C. E. Ford, a small study group was convened to attempt the formulation of a common system of nomenclature.

The meeting was arranged, through the good offices of Dr. T. T. Puck, to be held at Denver, in the University of Colorado, under the auspices of the Medical School. The meeting of this study group was made possible by the support of the American Cancer Society, to whom grateful thanks are due. For practical reasons, it was decided to keep the group as small as possible and to limit it to those human cytologists who had already published karyotypes.* In addition, three counsellors were invited to join the group to guide and aid the discussions and, if necessary, to arbitrate. Fortunately, the last office did not prove necessary, and it was possible by mutual agreement to arrive at a common system which has flexibility.

It was agreed that the principles to be observed by the system should be simplicity and freedom, as far as possible, from ambiguity and risks of confusion, especially with other systems of nomenclature in human genetics. It should also be capable of adjustment and expansion to meet the needs of new knowledge of human chromosomes. The system should be agreed to by the greatest possible proportion of cytologists working in the field, but the risk that a minority may be unable to accept the system as a whole should not be allowed to delay adoption by a majority.

It was agreed that the autosomes should be serially numbered 1 to 22 as nearly as possible in descending order of length, consistent with operational conveniences of identification by other criteria. The sex chromosomes should continue to be referred to as X and Y, rather than by a number which would be an additional and ultimately a superfluous appellation.

It was generally agreed that the 22 autosomes can be classified into seven groups, distinction between which can readily be made. Within these groups further identification of individual chromosomes can in many cases be made relatively easily. Within some groups, especially the group of chromosomes numbered 6–12, including also the X chromosome, the distinctions between the chromosomes are very difficult to make by presently available criteria. However, lesser difficulties are encountered in separating chromosomes 6 and the X from the remainder of this group. It is believed that, with very favourable preparations distinction can be made between most, if not all, chromosomes.

It is proposed that the autosomes first be ordered by placing the seven groups as nearly as possible in descending order of size. Within each group the chromosomes are arranged, for the most part, by size. It was desired specifically to avoid the implication that the size relationships have been permanently decided in every instance, but it is hoped that the assignment of numbers will be permanently fixed. In those cases where distinction is at present doubtful, final definition of each chromosome can be left until further knowledge has accrued, though an attempt is made to provide a characterisation of each. These principles make it possible to draw up a conspectus of the chromosomes, a table of their quantitative characteristics and a table of the synonyms which authors have already published. These are appended (tables I, II, and III).

In table II, showing the diagnostic characters of the chromosomes, three parameters are relied upon. These are:

(1) The length of each chromosome relative to the total length of a normal X-containing haploid set—i.e., the sum of the lengths of the 22 autosomes and of the X chromosome, expressed per thousand.

(2) The arm ratio of the chromosomes expressed as the length of the longer arm relative to the shorter one.

(3) The centromeric index expressed as the ratio of the length of the shorter arm to the whole length of the chromosome.

The two latter indices are, of course, related algebraically quite simply, but it is thought useful to present both here. In some chromosomes, the additional criterion of the presence of a satellite is available (table I), but in view of the apparent morphological variation of satellites, they and their connecting strands are excluded in computing the indices.

Table II shows the range of measurements determined by various workers. Some of the variation expresses the uncertainty due to measurement of relatively small objects; but many of the discrepancies between different workers' observations are due to the measurement of chromosomes at different stages of mitosis and to the effect of different methods of pretreatment and preparation for microscopic study. The ranges shown, therefore,

---

* In contemporary publications the terms, karyotype and idiogram, have often been used indiscriminately. We would recommend that the term, *karyotype*, should be applied to a systematised array of the chromosomes of a single cell prepared either by drawing or by photography, with the extension in meaning that the chromosomes of a single cell can typify the chromosomes of an individual or even a species. The term, *idiogram*, would then be reserved for the diagrammatic representation of a karyotype, which may be based on measurements of the chromosomes in several or many cells.

TABLE I—CONSPECTUS OF HUMAN MITOTIC CHROMOSOMES

Group 1–3	Large chromosomes with approximately median centromeres. The three chromosomes are readily distinguished from each other by size and centromere position.
Group 4–5	Large chromosomes with submedian centromeres. The two chromosomes are difficult to distinguish, but chromosome 4 is slightly longer.
Group 6–12	Medium-sized chromosomes with submedian centromeres. The X chromosome resembles the longer chromosomes in this group, especially chromosome 6, from which it is difficult to distinguish. This large group is the one which presents major difficulty in identification of individual chromosomes.
Group 13–15	Medium-sized chromosomes with nearly terminal centromeres (" acrocentric " chromosomes). Chromosome 13 has prominent satellites on the short arms. Chromosome 14 has small satellites on the short arms. No satellites have been detected on chromosome 15.
Group 16–18	Rather short chromosomes with approximately median (in chromosome 16) or submedian centromeres.
Group 19–20	Short chromosomes with approximately median centromeres.
Group 21–22	Very short, acrocentric chromosomes. Chromosome 21 has satellites on its short arms. The Y chromosome is similar to these chromosomes.

TABLE II—QUANTITATIVE CHARACTERISTICS OF THE HUMAN MITOTIC CHROMOSOMES

All measurements were made from cells of normal individuals, except those made by Fraccaro and Lindsten, which included cases of Turner's syndrome. Column A is the relative length of each chromosome, B is the arm ratio, and C the centromere index, as defined in the text.

	Tjio and Puck[6]			Chu and Giles[2]			Levan and Hsu[5]			Fraccaro and Lindsten *			Lejeune and Turpin[4] *			Buckton, Jacobs, and Harnden *			Range		
	A	B	C	A	B	C	A	B	C	A	B	C	A	B	C	A	B	C	A	B	C
1	90	1·1	48	90	1·1	48	85	1·1	49	82	1·1	48	87	1·1	48	83	1·1	48	82–90	1·1	48–49
2	82	1·6	39	83	1·5	40	79	1·6	38	77	1·5	40	84	1·5	40	79	1·6	38	77–84	1·5–1·6	38–40
3	70	1·2	45	72	1·2	46	69	1·2	45	65	1·2	45	67	1·2	46	63	1·2	46	63–72	1·2	45–46
4	64	2·9	26	63	2·9	26	63	2·7	27	62	2·6	28	62	2·6	25	60	2·6	28	60–64	2·6–2·9	25–28
5	58	3·2	24	58	3·2	24	59	2·6	28	60	2·4	29	57	2·4	30	57	2·4	30	57–60	2·4–3·2	24–30
X	59	1·9	34	57	2·8	38	52	1·6	38	54	1·6	38	58	2·2	32	51	1·7	37	51–59	1·6–2·8	32–38
6	55	1·7	37	56	1·8	36	54	1·6	38	54	1·6	38	56	1·7	37	56	1·6	38	54–56	1·6–1·8	36–38
7	47	1·3	43	52	1·9	35	51	1·9	35	50	1·7	37	51	1·8	36	50	1·7	37	47–52	1·3–1·9	35–43
8	44	1·5	29	46	1·7	29	48	1·6	33	47	1·7	37	48	2·4	29	46	1·5	40	44–48	1·5–2·4	29–40
9	44	1·9	40	46	2·4	38	47	1·8	36	45	2·0	33	47	1·9	35	44	2·1	32	44–47	1·8–2·4	32–40
10	43	2·4	27	45	2·3	30	45	2·0	33	45	2·6	34	45	2·6	27	44	1·9	35	43–45	1·9–2·6	27–35
11	43	2·8	34	44	2·1	32	44	2·2	31	43	2·2	31	43	2·5	40	43	1·5	40	43–44	1·5–2·8	31–40
12	42	3·1	24	43	3·1	24	42	1·7	32	43	1·7	37	42	2·8	27	42	2·1	32	42–43	1·7–3·1	24–37
13	35	8·0	11	32	9·7	10	32	5·0	16	34	4·8	17	33	6·8	14	36	4·9	17	32–36	4·8–9·7	10–17
14	32	7·3	12	34	9·5	9	37	4·0	18	35	4·4	19	32	7·0	13	34	4·3	19	32–37	4·3–9·5	9–19
15	29	10·5	9	31	11·9	8	35	4·7	17	33	4·6	22	31	10·0	9	34	3·8	22	29–35	3·8–11·9	8–22
16	32	1·8	36	27	1·6	38	30	1·4	42	31	1·4	42	29	1·4	41	33	1·4	31	27–33	1·4–1·8	31–42
17	29	2·8	26	30	2·1	33	29	2·4	30	30	1·9	35	29	3·1	23	30	1·8	36	29–30	1·8–3·1	23–36
18	24	3·8	21	25	3·8	22	25	2·6	28	27	2·5	29	27	2·4	21	27	2·4	29	24–27	2·4–4·2	21–29
19	22	1·4	41	22	1·9	34	24	1·2	40	25	1·3	43	22	1·4	42	26	1·2	45	22–26	1·2–1·9	34–45
20	21	1·3	44	19	1·3	44	21	1·2	40	23	1·3	43	20	1·2	43	25	1·2	46	19–25	1·2–1·3	40–46
21	18	3·7	21	15	6·8	13	13	2·5	28	19	2·5	29	15	2·3	31	20	2·5	29	13–20	2·3–6·8	13–31
22	17	3·3	23	12	6·0	14	16	2·0	33	17	2·3	30	13	4·0	20	18	2·7	27	12–18	2·0–6·0	14–33
Y	19	∞	0	11	∞	0	18	4·9	17	22	2·9	26	18	∞	0	18	4·9	17	11–22	2·9–∞	0–26

* Unpublished data.

represent the maxima and minima of the means found by different workers using different techniques. However, within any one worker's observations, the variations are not so broad.

Reference should be made to two other matters of nomenclature. In the first place, it is considered that no separate nomenclature for the groups is needed. It is considered that any group to which it may be necessary to refer will be a sequence of those designated by Arabic numerals. Hence, any chromosome group may be referred to by the Arabic numerals of the extreme chromosomes of the group, joined together by a hyphen—e.g., the group of the three longest chromosomes would be Group 1–3. This scheme has the merit of great flexibility. For instance, chromosomes X and 6 may be separated from the Group 6–12 whenever they can be distinguished.

Secondly, there is the problem raised by the abnormal chromosomes which are being encountered in the more recent studies. Their nomenclature was discussed without a definite conclusion being reached. Broadly, it was agreed, however, that any symbol used should avoid incorporating a specific interpretation which was not reasonably established. It was suggested that arbitrary symbols, prefixed by a designation of the laboratory of origin, should usually be assigned to the abnormal chromosome.

In this connection, two further requisites for coordination of research were discussed. One is the storage of documentation for reference, perhaps in a central depository, additional to what it may be possible to publish. The other is the desirability that cultures be preserved, by the satisfactory methods now used, so that they are available for reference, comparison, and exchange.

Some consideration was also given to the desirability of using a uniform system for presenting karyotypes and idiograms, but recognising that individual variation in taste is involved, rigidity of design was thought undesirable. However, it was recommended that the chromosomes should be arranged in numerical order, with the sex chromosomes near to but separated from the autosomes they resemble. It is desirable that similar ones be grouped together with their centromeres aligned.

It is recognised that choice between the different possible schemes of nomenclature is arbitrary, but that uniformity for ease of reference is essential. Hence, individual preferences have been subordinated to the

TABLE III—SYNONYMY OF CHROMOSOMES AS PUBLISHED BY VARIOUS WORKERS

New chromosome number	Tjio and Puck[6]	Chu and Giles[2]	Levan and Hsu[5]	Ford, Jacobs and Lajtha[3]	Böök, Fraccaro, and Lindsten[1]	Lejeune, Turpin, and Gautier[4]
1	1	1	1	1	1	G1
2	2	2	2	2	2	G2
3	3	3	3	3	3	G3
4	4	4	4	4	4	G4
5	5	5	5	5	5	G5
6	6	6	6	6*	6	M1
7	7	7	7	7	7	M2
8	8	8	8	(8)	8	Md1
9	9	9	9	(9)	9	M3
10	10	10	10	(11)	10	Md2
11	11	11	11	(12)	11	M4
12	12	12	12	(13)	12	Md3
13	18	14	20	14	14	T1
14	19	15	18	18	15	T2
15	20	13	19	19	13	T3
16	13	17	15	15	19	C1
17	14	16	13	17	17	P1
18	15	18	14	18	18	P2
19	16	19	16	20	19	C2
20	17	20	17	21	20	C3
21	21	21	21	22	21	Vh
22	22	22	22	23	22	Vs
X	X	X	X	?(7)	X	X
Y	Y	Y	Y	Y	Y	Y

* In the published idiogram the chromosomes of group 6–12 (including X) were indicated by discontinuous lines and left unnumbered owing to the uncertainty of discrimination at that time. For the purpose of this table, these chromosomes have been assigned the numbers shown in parentheses, in serial order of length.

1. Böök, J. A., Fraccaro, M., Lindsten, J. Acta pædiat. 1959, 48, 453.
2. Chu, E. H. Y., Giles, N. H. Amer. J. hum. Genet. 1959, 11, 63.
3. Ford, C. E., Jacobs, P. A., Lajtha, L. Nature, Lond. 1958, 181, 1565.
4. Lejeune, J., Turpin, R., Gautier, M. Ann. Génét. 1959, 2, 41.
5. Levan, A., Hsu, T. C. Hereditas, 1959, 45, 665.
6. Tjio, J. H., Puck, T. T. Proc. Nat. Acad. Sci. 1958, 44, 1229.

common good in reaching this agreement. This human chromosomes study group therefore agrees to use this notation and recommends that any who prefer to use any other scheme should, at the same time, also refer to the Standard System here proposed.

We are well aware of the wide interest in the work of this study group and realise that this meeting is merely a preliminary to a larger meeting. It is believed that two needs have to be met in this respect. One is for seminars and workshops at which workers in the field may exchange information; such seminars are best arranged regionally. The second need, which may come later, is for international conferences; and we believe that congresses and other organisations whose interests include human genetics, should promote such meetings.

Participants

J. A. BÖÖK
Institute for Medical Genetics, Uppsala.

J. LEJEUNE
Hospital Trousseau, Paris.

A. LEVAN
Institute of Genetics, Lund.

E. H. Y. CHU
Oak Ridge National Laboratory, Tennessee.

C. E. FORD
M.R.C. Radiobiological Research Unit, Harwell, Berks.

M. FRACCARO
Institute for Medical Genetics, Uppsala.

D. G. HARNDEN
M.R.C. Group for Research on the General Effects of Radiation, Western General Hospital, Edinburgh.

T. C. HSU
M. D. Anderson Hospital and Tumor Institute, Houston, Texas.

D. A. HUNGERFORD
Institute for Cancer Research, Philadelphia.

P. A. JACOBS
M.R.C. Group for Research on the General Effects of Radiation, Western General Hospital, Edinburgh.

S. MAKINO
Zoological Institute, Hokkaido University, Sapporo, Japan.

THEODORE T. PUCK
University of Colorado Medical Center, Denver.

A. ROBINSON (secretary)
Department of Biophysics, University of Colorado Medical Center, 4200 East Ninth Avenue, Denver 20 Colorado, U.S.A.

J. H. TJIO
National Institutes of Health, Bethesda, Maryland.

Counsellors

D. G. CATCHESIDE (chairman)
Department of Microbiology, The University, Edgbaston, Birmingham 15, England.

H. J. MULLER
Indiana University, Bloomington, Indiana.

CURT STERN
University of California, Berkeley, California.

CHAPTER 8

Later years: the growth of human and clinical cytogenetics

THIS BOOK HAS CONCENTRATED ON the work – and workers – of the 'first years' of modern human cytogenetics, the period from 1955 to 1960. That it has not attempted to cover the subsequent growth and development of the field in any depth does not indicate any natural break after 1960, but simply the practical constraints mentioned in the introduction. This growth was so rapid, with the scientists involved soon becoming a sizeable body rather than the original handful, that it would be impossible to do them or the field justice, or to give the personal backgrounds that I have tried to supply here for these initial years. As made clear at the beginning of the book, this is not intended to be a comprehensive history of human cytogenetics, something that remains to be written, though some other works, such as those of Harris[1] and Hsu[2] have partly covered the topic.

To leave the reader 'suspended' at 1960, though, with no indication as to the broad directions taken by the field during the next 40 years, would be unhelpful, especially as these later developments were often powerfully influenced by events up to 1960 and also resolved many of the problems still unsolved at this point. This chapter therefore tries to provide a link between the beginnings and the present, but necessarily can only do so in outline and selectively. I have tried where possible to give references to some historical reviews and personal accounts, but have not included interview excerpts here, even though I have been fortunate enough to have had discussions with a number of the key people involved, who of course include many of those who have already figured in earlier chapters. I have given some brief notes also on some of the principal workers not mentioned previously.

New chromosomal syndromes, 1961–1970

Even though the quality of human chromosome preparations had not changed greatly since the quantum leap given by Tjio's studies, the possibility of using blood and the various simplifications of technique described in Chapter 7 meant that rapidly increasing numbers of samples from a wide range of medical situations were being analysed, notably from children with congenital abnormalities; this compensated for the relative rarity of abnormal chromosome complements found by early investigators.

Table 8-1 summarises the main discoveries of autosomal abnormalities in liveborn children over this decade; not surprisingly, these abnormalities were mostly substantial deletions of arms of chromosomes, such as 4p, 5p, and 18q, or loss of an entire small chromosome, such as 21 monosomy, which could be reliably detected now that the measurements of the different chromosomes (and their nomenclature) had been accurately established and standardised. A wide range of sex chromosome anomalies was also recognised during this decade, including isochromosomes for the X chromosome, the XYY karyotype (mentioned in Chapter 4) and various combinations of multiple X and Y chromosomes, often combining a Klinefelter phenotype with significant mental retardation.

Table 8-1 also shows the groups involved and emphasises the importance of contributions from France during this period. Some of the first scientists involved in human cytogenetics, including Tjio, Levan, Hsu and Ford, had returned to their primary interests of cancer research and basic science, but others stayed with the new field of clinical cytogenetics; notably Lejeune, but also Patricia Jacobs and John Hamerton, the latter at this point at Guy's Hospital, London with Paul Polani and later in Winnipeg, Canada. Both Lejeune and Hamerton had natural bases in academic paediatric units for their work, but the Edinburgh MRC unit (unlike its counterpart at Harwell) was also able to broaden its research remit to include population cytogenetics, particularly of the sex chromosome abnormalities, as well as continuing its radiation and leukaemia research.

TABLE 8-1
NEW CHROMOSOMAL SYNDROMES, 1961–1970

Abnormality	Researcher
4p– (Wolf–Hirschhorn syndrome)	Wolf *et al.*, 1965; Hirschhorn *et al.*, 1965
5p– (cri du chat syndrome)	Lejeune *et al.*, 1963
5p trisomy	Lejeune *et al.*, 1965
13q monosomy	Lejeune *et al.*, 1968
18p– syndrome	de Grouchy *et al.*, 1963
18q– syndrome	de Grouchy *et al.*, 1964
21 partial monosomy	Lejeune *et al.*, 1964

Fig. 8-1 Jean de Grouchy, 1926–2003 (courtesy of Professor Catherine Turleau).

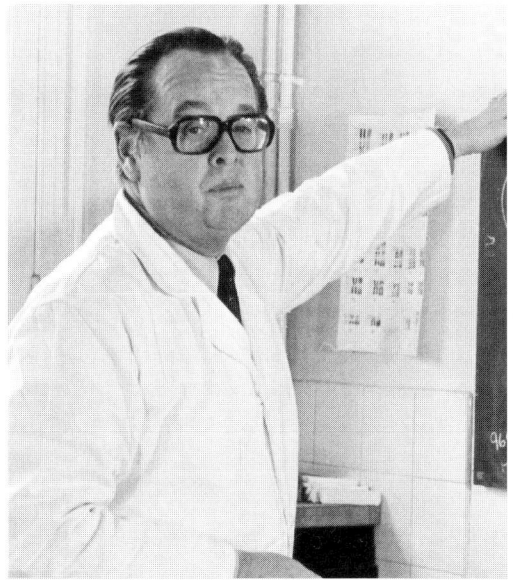

Fig. 8-2 Kurt Hirschhorn (courtesy of Dr Kurt Hirschhorn).

Lejeune's group was not the only one in Paris to contribute to this new series of discoveries. Jean de Grouchy, a close personal friend of Lejeune, developed a laboratory for clinical and comparative cytogenetics under Maurice Lamy at Hôpital Necker, and after Lejeune had moved there with Raymond Turpin in 1968, this hospital became one of the major world centres of clinical cytogenetics (Figure 8-1). The contributions of these two closely linked groups are epitomised by their two classic books: Turpin and Lejeune's *Les Chromosomes Humains* (translated as *Human Affliction and Chromosome Aberration*)[3] and the *Clinical Atlas of Human Chromosomes*, by de Grouchy and Catherine Turleau.[4]

American groups, too, were now beginning to make a wider impact on clinical cytogenetics, after a surprisingly late start. As we have seen, only the laboratories of Klaus Patau and Eeva Therman in Madison, Malcolm Ferguson-Smith (temporarily in Baltimore from Britain between 1959 and 1962) and David Hungerford in Philadelphia in the leukaemia field, had made significant contributions up to 1960. Kurt Hirschhorn's (Figure 8-2) identification of the 5p– syndrome[5] (the full paper was published simultaneously with that of Wolf and colleagues in Germany,[6] but an abstract had been published already in 1960) gave foundations for the field in New York from 1960 onwards, as did the work of Orlando J Miller. As mentioned previously, both of these had spent time at the Galton Laboratory and at Uppsala before returning to America and both have written reviews of the early human cytogenetics research. Their work, along with that of James German and Harold Klinger, also in New York, and of Irene Uchida in Winnipeg, Canada, who had worked with Klaus Patau, rapidly radiated out across the rest of North America.

Chromosomes and spontaneous abortion

The first papers on trisomies 18 and 13 had both suggested that trisomy for the larger chromosomes might not be viable, while Penrose and Delhanty's finding of triploidy in a stillborn infant provided an example of this.[7] However, it was the contribution of David Carr, working in the anatomy department of Murray Barr in London, Ontario, that showed definitively the importance and frequency of chromosomal abnormalities in relation to spontaneous abortion. In a preliminary paper in 1963,[8] he reported finding such abnormalities in 12 of 53 abortions, but it was his comparable findings in a much larger series (44 out of 200, 22 per cent) given in a detailed paper in 1965[9] that had most impact, largely because of the type of chromosome abnormality found, not just on account of the high frequency.

Carr indeed found some of the predicted non-viable trisomies of larger chromosomes in his series, while triploidy was the second most common chromosome anomaly to be found, with three examples; but other findings were less expected, notably that the commonest abnormality was the 45X Turner karyotype (11 out of the total of 44 abnormalities), an observation contrasting with its relatively mild phenotype in living patients and its low frequency in adult and paediatric cytogenetic series. Other sex chromosome abnormalities were also commoner than predicted from their adult frequency, and so were trisomy 21 and the D and G group trisomies.

It was thus clear from Carr's results that there was no discontinuity between those chromosome abnormalities found in abortions and those in liveborns, but that most showed a gradient, with liveborns representing a minority of cases, and with the conception frequency of chromosome defects an order of magnitude higher than previously thought likely, representing a major biological phenomenon as well as an important medical problem. Carr's work set the stage for population cytogenetic studies of pregnancy loss by Patricia Jacobs[10] and others, as well as for the identification of chromosomal factors in recurrent abortion and infertility. Equally important was that it focused the attention of obstetricians on chromosomes and on genetics generally, which now provided a practical diagnostic tool in investigating their patients.

X chromosome biology

It was mentioned at the end of Chapter 4 that the discovery of the sex chromosome abnormalities provided the foundation for more general and fundamental research on the biology of the sex chromosomes. This is far too extensive a field to do more than touch on here, and it extends outside cytogenetics, but one or two key discoveries must be mentioned. Ursula Mittwoch's 1967 and 1973 monographs provide valuable accounts of this work.[11,12]

The topic most directly cytogenetic in nature is the pairing mechanism of the human sex chromosomes and the long debated question of whether the X and Y chromosomes indeed had a homologous region. This was conclusively shown to be the case by the observations of Ferguson-Smith[13] and has since been confirmed in detail by molecular studies, with some XX males showing a translocated portion of Y chromosome material.

The most fundamental advance, though, arising directly from the original sex chromatin work, relates to X chromosome inactivation. It was not until successive X chromosome abnormalities had been identified that it became clear that the number of sex chromatin bodies was always one less than the number of X chromosomes, with each body representing one X chromosome. The concept that the sex chromatin represented a condensed and inactive X, put forward by Ohno and others and reviewed by Ohno,[14] then led to

Fig. 8-3 Mary Lyon (courtesy of Dr Mary Lyon and reprinted, with permission, from *Annual Review of Genetics*, Volume 26 © 1992 by Annual Reviews, www.annualreviews.org).

Mary Lyon's 1961 hypothesis of random X inactivation,[15] which has become one of the cornerstones of mammalian biology (Figure 8-3).

Although put forward on the basis of coat colour observations in the mouse, its general application was immediately apparent, notably explaining the heterozygous variability of many human X-linked disorders, as shown by Beutler,[16] and by Lyon herself.[17]

Chromosome banding

By the end of the 1960s the existing chromosome technology had again reached its limits and momentum was beginning to be lost. It was still not possible to identify all human chromosomes individually with certainty, at least not by using easily applicable methods, and internal rearrangements or exchanges of chromosome material were mostly undetectable. This situation was changed completely by the development of chromosome banding techniques in the laboratory of Torbjörn Caspersson at the Karolinska Institute, Stockholm (Figure 8-4), in a series of publications between 1969 and 1971.[18-20]

While Caspersson's name is generally associated with this discovery, the work was in fact largely developed by his colleague Lore Zech (Figure 8-5), and Caspersson himself was not only little involved, but was reluctant to be convinced by the relevance or even existence of the bands, at least for human chromosomes (interview with Professor Lore Zech in September 2004). From all accounts Caspersson was a far from easy person to work with; his policy of insisting on being first author on all his papers (a policy also incorrectly ascribed to Albert Levan), led to tensions, as did his unwillingness to collaborate with other groups, even in the same institute. Despite this, Lore Zech describes his department, with its high international profile, as an exciting place to work in.

Chromosome bands were actually far from being a new observation (even in the human chromosome preparations of Tjio faint banding can be seen), but the devel-

Fig. 8-4 Torbjörn Caspersson (Photograph reproduced from *The Cells of the Body* (1995) by Henry Harris, © Cold Spring Harbor Laboratory Press, with permission.)

Fig. 8-5 Lore Zech (courtesy of Professor Lore Zech).

opment of a series of fluorescent stains, especially those not fading rapidly, like quinacrine mustard, showed that the banding pattern could provide the basis for unequivocal identification of all human chromosomes, including the Y chromosome, and potentially for detecting abnormalities resulting in rearrangement of the bands. Variant techniques, such as R (reverse) and C (centromeric) banding extended the applications.[21, 22]

As so often is the case with new technology, this advance initially required complex and cumbersome equipment (according to Lore Zech this was also accentuated by Caspersson's fascination with large machines). Soon, though, the discovery that Giemsa staining could give the same banding pattern as quinacrine fluorescence, and the publication by Marina Seabright of a simple method based on this,[23] allowed chromosome banding to become part of routine clinical cytogenetics.

A wide range of scientific discoveries and clinical applications rapidly arose from the use of the new banding methods. Especially notable was the recognition of numerous reciprocal translocations, balanced and unbalanced, where the rearrangement, and loss or addition of material, would have previously been undetectable. These proved important not only in familial malformations, spontaneous abortions and infertility, but they allowed Janet Rowley (Figure 8-6) to show that the Ph^1 chromosome in chronic

Fig. 8-6 Janet Rowley (courtesy of Dr Janet Rowley).

myeloid leukaemia was indeed the result of a translocation, by demonstrating the presence of the missing portion of chromosome 22 on chromosome 9q.[24] Likewise, a specific translocation was found by Lore Zech and colleagues to be the hallmark of the Burkitt lymphoma.[25]

Previously unrecognisable deletions provided the basis for a series of further malformation syndromes, progressively extended to 'microdeletion syndromes' as it became possible to study less condensed chromosomes. Again, this had an important impact on cancer genetics by showing that familial embryonic tumours, notably retinoblastoma, could be associated with the loss of specific constitutional chromosome material,[26] while the previously confusing studies of the tumours themselves could now in some cases be sorted into abnormalities of specific chromosome regions. An unexpected result, important theoretically as well as practically, was the finding that the Y chromosome could be identified by its fluorescence not only in dividing but in interphase cells.[27]

All of this new detail of chromosome morphology provided a challenge for those responsible for nomenclature, now the responsibility of a specific body, International System for Human Cytogenetic Nomenclature (ISCN), as indicated in Chapter 7. The original Denver nomenclature was extended, without too many problems, at the 1971 international Paris Congress,[28] and the resulting ISCN classification provided the definitive basis for how we represent chromosomes at present.

Thus, the discovery of chromosome banding produced a further decade of new cytogenetic discoveries, until by 1980 the stage was set for the fusion of cytogenetics with the new molecular techniques that were beginning to have applications in human and mammalian genetics.

Cytogenetic diagnostic laboratories

Almost all the work described in this book was undertaken as part of original research, by scientists for whom this, rather than clinical applications was the main aim. However helpful and necessary the close links with clinicians were in allowing these advances, the point was rapidly reached when diagnostic genetic services needed to be established and funded in their own right, rather than on the back of research activities.

This process happened very rapidly, on a worldwide basis, during the early and mid-1960s; the process of transition, approached differently in different centres and countries, would make an interesting historical study in its own right. In most European countries with systems of universal health care, chromosome analysis was incorporated into these systems; continuing links with research, though, allowed more basic work to benefit from the growing number of newly recognised chromosome abnormalities, while service laboratories were able to rapidly adopt and modify technological advances such as banding.

The recognition in 1966 that cultured amniotic fluid cells (and subsequently chorion villus biopsy) could permit prenatal diagnosis of human chromosome disorders,[29] following the use of sex chromatin analysis on amniotic fluid cells as long ago as 1956,[30] provided a further source of clinical demand for diagnostic cytogenetics laboratories, albeit one raising major ethical as well as technical challenges. This is another area that is ready for a detailed historical study.

Chromosomes and gene mapping

Perhaps the most important scientific role of chromosome banding techniques was to give a detailed physical basis for the rapidly developing human gene map, especially when combined with the use of somatic cell hybrids. Since the completion of the international human genome project it has become easy to think of this exclusively in terms of large-scale DNA sequencing, but it was cytogenetic analysis that was largely responsible for providing the framework on to which most of the sequencing effort could be fitted. In particular the mapping and cloning of many important disease genes was founded on the observation of key patients showing both the disease and a specific chromosome defect, usually a deletion or translocation. Retinoblastoma and Duchenne muscular dystrophy are cases in point. The detailed ordering of genes frequently depended on analysis of natural or experimental cell lines showing a series of deletions of a particular chromosomal region.

Normal variation in chromosome morphology could also be exploited in gene mapping and the first human gene assignment to a specific autosome was achieved in this way – the Duffy blood group locus to the 'uncoiled' region of chromosome 1.[31]

A powerful experimental approach to human gene mapping proved to be the development of interspecific hybrid cell lines[32] where the human chromosomes were lost in a progressive and orderly manner, allowing the correlation of presence or absence of a particular characteristic, often an enzyme, with a specific human chromosome or chromosome region.

Chapter 1 made the point that it was a combination of cytogenetics and classical genetic linkage analysis that established the original gene maps of *Drosophila*. The same combination, together with information from genetic diseases, was responsible for the detailed human gene map that was already well established before total sequencing became a possibility.

The fusion of cytogenetics and molecular genetics: *in situ* hybridisation

As long ago as the early 1970s attempts were made to identify and locate specific nucleic acid sequences visually by various approaches to *in situ* hybridisation, particularly by using autoradiography, which had been applied to human chromosomes since the early 1960s, to detect variation in radioactive thymidine uptake.[33] Success was limited, though, to the highly repetitive DNA sequences and their corresponding RNA in the satellited regions of chromosomes,[34] and to the centromeres. The situation changed radically once recombinant DNA techniques could produce numerous molecules of single copy DNA sequences, the ones of most relevance to science and medicine as representing specific genes. As with chromosome banding methods, it was the development of fluorescent molecules that could bind, directly or via an intermediate molecule, to DNA bases, which led to the powerful technique of fluorescent *in situ* hybridisation (FISH), which has now firmly linked cytogenetics to molecular genetics.[35] Again, like the banding development, this has opened up a new range of applications that have given the microscopic basis of classical cytogenetics a new lease of life, especially in the medical diagnostics field, but also in gene ordering and mapping. Malcolm Ferguson-Smith, himself one of the main contributors to this work, has provided a valuable historical account of the development of these and earlier cytogenetic methods in relation to human gene mapping.[36]

Nor have FISH techniques been limited to single gene abnormalities. They have powerfully extended the range of microdeletions and other small chromosome alterations that can be recognised, bridging the gap between those visible by conventional banding techniques and defects at the single gene level. Likewise, the use of 'chromosome paints' based on DNA from sorted chromosomes can allow the chromosome origin of complex rearrangements, such as those often seen in cancer cells, to be clearly recognised. How the early workers on cancer chromosomes, such as Albert Levan, would have appreciated such techniques in their own pioneering work!

References

1. Harris H (1995). *The Cells of the Body. A History of Somatic Cell Genetics*. Cold Spring Harbor, CSHL Press.
2. Hsu TC (1979). *Human and Mammalian Cytogenetics. An Historical Perspective*. New York, Springer.
3. Turpin R and Lejeune J (1969). *Human Afflictions and Chromosomal Aberrations*. Oxford, Pergamon Press. (Original French edition *Les Chromosomes Humains* published in 1965).
4. de Grouchy J and Turleau C (1984). *Clinical Atlas of Human Chromosomes* (*Atlas des Maladies Chromosomiques*). New York, Wiley.
5. Hirschhorn K, Cooper HL and Firschein I (1965). Deletion of short arms of chromosome 4–5 in a child with defects of midline fusion. *Humangenetik* **1**, 479–482.
6. Wolf U, Reinwein H, Porsch R, Schroter R and Baitsch H (1965). Deficienz an den kurzen armen eines chromosoms nr 4. *Humangenetik* **1**, 397–413.
7. Penrose LS and Delhanty JDA (1961). Triploid cell cultures from a macerated foetus. *Lancet* **1**, 1261–1262.
8. Carr DH (1963). Chromosome studies in abortuses and stillborn infants. *Lancet* **2**, 603.
9. Carr DH (1965). Chromosome studies in spontaneous abortions. *Obstet. Gynecol.* **26**, 306–326.
10. Hassold TJ, Matsuyama A, Newlands IM, Matsuura JS, Jacobs PA, Manuel B and Tsuei B (1978). A cytogenetic study of spontaneous abortions in Hawaii. *Ann. Hum. Genet.* **41**, 443–454.
11. Mittwoch U (1967). *Sex Chromosomes*. New York, Academic Press.
12. Mittwoch U (1973). *Genetics of Sex Differentiation*. New York, Academic Press.
13. Ferguson-Smith MA (1965). Karyotype–phenotype correlations in gonadal dysgenesis and their bearing on the pathogenesis of malformations. *J. Med. Genet.* **2**, 142–155.
14. Ohno S (1966). Single-X derivation of sex chromatin. In: Moore KL (ed.), *The Sex Chromatin*. Philadelphia, WB Saunders, pp. 113–128.
15. Lyon MF (1961). Gene action in the X-chromosome of the mouse (*Mus musculus* L.). *Nature* **190**, 372–373.
16. Beutler E, Yeh M and Fairbanks VF (1962). The normal human female as a mosaic of X-chromosome activity. *PNAS* **48**, 9–16.
17. Lyon MF (1962). Sex chromatin and gene action in the mammalian X-chromosome. *Am. J. Hum. Genet.* **14**, 135–148.
18. Caspersson T, Zech L, Modest EJ, Foley GE, Wagh U and Simonsson E (1969). Chemical differentiation with fluorescent alkylating agents in Vicia faba metaphase chromosomes. *Exp. Cell Res.* **58**, 128–140.
19. Caspersson T, Zech L and Johansson C (1970).

19. Differential binding of alkylating fluorochromes in human chromosomes. *Exp. Cell Res.* **60**, 315–319.
20. Caspersson T, Lomakka G and Zech L (1971). The 24 fluorescence patterns of the human metaphase chromosomes – distinguishing characters and variability. *Hereditas* **67**, 89–102.
21. Dutrillaux B and Lejeune J (1971). Sur une nouvelle technique d'analyse du caryotype humain. *C. R. Acad. Sci. Paris* **272**, 2638–2640.
22. Arrighi FE and Hsu TC (1971). Localization of heterochromatin in human chromosomes. *Cytogenetics* **10**, 81–86.
23. Seabright M (1971). A rapid banding technique for human chromosomes. *Lancet* **2**, 971–972.
24. Rowley JD (1973). A new consistent chromosomal abnormality in chronic myelogenous leukemia identified by quinacrine fluorescence and giemsa staining. *Nature* **243**, 290–293.
25. Zech L, Haglund U, Nilsson K and Klein G (1976). Characteristic chromosomal abnormalities in biopsies and lymphoid cell lines from patients with Burkitt and non-Burkitt lymphoma. *Int. J. Cancer* **17**, 47–56.
26. Francke U (1976). Retinoblastoma and chromosome 13. *Cytogenet. Cell Genet.* **16**, 131–134.
27. Pearson PL and Bobrow M (1970). Technique for identifying Y chromosomes in human interphase nuclei. *Nature* **226**, 78–80.
28. Paris Conference (1971). *Standardization in Human Cytogenetics. Birth Defects Original Articles Series*, vol 8, no 7. National Foundation, New York.
29. Steele MW and Breg WR (1966). Chromosome analysis of human amniotic fluid cells. *Lancet* **1**, 383–385.
30. Fuchs F and Riis P (1956). Antenatal sex determination. *Nature* **177**, 330.
31. Donahue RP, Bias WB, Renwick JH and McKusick VA (1968). Probable assignment of the Duffy blood group locus to chromosome 1 in man. *PNAS* **61**, 950–955.
32. Weiss MC and Green H (1967). Human-mouse hybrid cell lines containing partial complements of human chromosomes and functioning human genes. *PNAS* **58**, 1104–1110.
33. German JL and Bearn AG (1961). Asynchronous thymidine uptake by human chromosomes. *J. Clin. Invest.* **40**, 1041–1042.
34. Pardue ML and Gall JG (1970). Chromosomal localization of mouse satellite DNA. *Science* **168**, 1356–1358.
35. Fan Y-S, Davis LM and Shows TB (1990). Mapping small DNA sequences by fluorescence *in situ* hybridization directly on banded metaphase chromosomes. *PNAS* **87**, 6223–6227.
36. Ferguson-Smith MA (1993). From chromosome number to chromosome map: the contribution of human cytogenetics to genome mapping. *Chromosomes Today* **11**, 3–19.

CHAPTER 9

Conclusions

WHAT CAN WE LEARN FROM THE fascinating period of scientific development that I have tried to describe in this book? The scientific story speaks for itself: the study of human chromosomes moved from being an area that was largely ignored and had no seeming clinical significance in 1955, to being one of the most exciting fields of biology and medicine, until by 1960, it was on the threshold of becoming an important medical laboratory discipline in its own right, and having a powerful influence on allied fields such as clinical genetics and paediatrics.

I believe, though, that these 'first years of human chromosomes' can teach us about more general factors that are important in the development of science and medicine, factors that are often more readily learned from conversations with those actually responsible for the work than from the final, more polished, published papers. I try to outline some of these factors below.

Excitement and enjoyment

It has struck me forcibly, during my interviews and conversations, how much those involved enjoyed their work. This enjoyment seemed as strong as ever 50 years later and I hope that some of this has come across in the quotations that I have included, especially in the excerpts from the recordings. The 2004 description by Mike Bertram and Keith Moore of the sex chromatin discovery might have been about something that happened yesterday, not in 1949, and much the same applies to the other descriptions including, significantly, that of Muriel Lee, working as a technician with Patricia Jacobs. All seem to be have been infected with the excitement of forming part of the new field opening up around them and to have had a real devotion to their work, often lifelong, as shown by Albert Levan's statement that 'after looking at chromosomes every day for 50 years, I regard them as my friends'.

The role of individuals

In an age of 'big science', where most advances are now made by large teams and where it is often difficult to work out whose role was most significant, these early discoveries in the field of human chromosomes were made by individuals, often collaborating in groups of two or three, but with the work almost always attributable to specific individuals; this again is seen most clearly in the points of detail and background mentioned in the interviews.

Harris, in the introduction to his book *The Cells of the Body*, mentions how he found reading the work of 19th century authors more interesting than that by modern large teams. Perhaps this is partly because of the more personal style these early authors used – and the greater space that they were allowed by journals! We can still see this reflected in the spirited style of the early papers of Winiwarter, but by the 1950s published papers had become more condensed and less personal, so that much of the subjective 'character' of the individual authors involved has been lost. To find this one needs to turn again to the interviews and recordings, as well as to personal reminiscences and recollections, published or unpublished, such as those of TC Hsu, Maj Hultén or Paul Polani. Without such personal accounts we would not have such vivid pictures as those of Joe Hin Tjio working through the night and sleeping under the laboratory bench (in a sleeping bag borrowed from Maj Hultén!), or of Muriel Lee's glee when Patricia Jacobs correctly identified the additional set of XXY slides that, unknown to her, Muriel had slipped in without saying anything.

Independence and youth

It is striking how large a part in these advances was played by scientists who were relatively junior and often very young. David Hungerford was a postgraduate student, David Harnden in his first post-doctoral position, while Patricia Jacobs was aged just 23, at the time of their major discoveries, but all were given a large degree of independence and responsibility for their work. One senses that 'supervision', where it existed, was very much a background affair and that there was freedom to develop the work as seemed natural and productive. Likewise, the primary workers were also the primary authors; heads of department (where they figure at all) do not seem to have 'taken over' the credit for the work, nor attempted to do so. The example of Levan sometimes quoted in this respect is clearly incorrect, quite apart from the fact that he was not head of the department. The role of Michael Court Brown in supporting, encouraging and co-ordinating the work of the Edinburgh group (and in helping with the writing of papers; see Chapter 4) comes across as a model of how helpful an experienced and involved unit head can be.

The key role of women as scientists in cytogenetics

At the time of the work covered by this book, mostly 40–50 years ago, it was far from easy for a woman to be able to develop a career as an independent worker, and correspondingly far too frequent for male colleagues to assume an undue proportion of the credit for any major discovery. During the course of the interviews underpinning this book I have been struck by how many of the important contributions have been made by women, not all of whom have, at least initially, received as full recognition as they deserved. A telling example can be seen in the printed discussion following the third paper of Lejeune, Turpin and Gautier (see Chapter 3, reference 14), where the chairman is recorded as pointedly congratulating 'Messieurs Lejeune and Turpin', while making no mention of Marthe Gautier! I hope that in this account I have given a fair balance and have been able to emphasise some key roles that are less well known than they should be.

Collaboration between scientists and clinicians

Once the development of basic techniques and the unambiguous determination of the human chromosome number had suggested the possibility of medical applications of human cytogenetics, the need for collaboration between the two very different worlds of basic science and clinical medicine suddenly became essential. The gap here was probably wider than in many other areas of medical science because, apart from the cancer field, there had previously been few reasons relating to chromosomes for either basic scientists or clinicians to seek each other's help.

For the most part in the work described here, these collaborations seem to have been both fruitful and harmonious, with those involved each recognising and respecting the other's skills and role. We are probably seeing a biased selection here though, because investigators without such links, or where collaborations did not work out, would have been largely unsuccessful. An important factor in this early phase was the involvement of key linking individuals who had both clinical and genetic expertise (e.g., Paul Polani, John Edwards) and who were thus able to play a particularly active role in initiating and focusing the collaborations.

Flexibility of institutions and funding bodies

It is easy to forget that most of the British work on human sex chromosome anomalies and congenital malformations was done by workers who were supposed to be focusing on radiation-related leukaemia or on mammalian and radiation cytogenetics. This is a reminder of the important impetus that radiation effects had given to human chromosome research, but it is remarkable that bodies like the UK Medical Research Council (MRC) do not seem to have objected to their workers straying into fields quite different from their primary remit! It would be interesting to see if official records contain any doubts or complaints about this, but those scientists involved, when I spoke with them, are definite that they received only encouragement and met no bureaucratic obstacles, although it is quite possible that the directors of their units (Michael Court Brown in Edinburgh, John Loutit at Harwell) did some protective liaison on their behalf. It must also be remembered that the 'bureaucracy' was much slimmer then, with just one or two, often talented, administrators responsible for wide areas of research, and with minimal paperwork involved.

A similar attitude can be seen in Philadelphia, where the institutions involved were happy to support not only Nowell's 'unfocused research on leukaemia', but his initially informal collaboration 'across the road' with David Hungerford, whose official research was on a quite different topic. Wisely, these institutions seem to have taken the view that their role was to appoint talented workers, support them and then to leave them undisturbed.

Would such flexibility be possible now? I have doubts, at least in terms of official backing; although investigators have become adept at initiating new and promising research 'on the back of' funding given for something else, any longer-lasting programme would now need to be more specifically justified.

A small and interactive community

A striking feature of these years is how few people and centres were involved in the initial discoveries. Even by the time of the 1960 Denver conference, where invitations were limited to those who had published a human karyotype, there were only 14 such individuals, from nine centres worldwide, a number that would soon rise exponentially. Such meetings, and the fact that all those involved could fit into a small room, must have greatly increased the sense of community, despite the difficulties encountered in reaching agreement on nomenclature mentioned in Chapter 7, and the strongly held views and individualistic characters of many of the workers involved.

Inevitably, such closeness and informality became increasingly difficult to maintain with the rapid growth of the fields after 1960, but regular meetings such as the 'Oxford (later international) chromosome conferences', started in 1964, helped to maintain links.

Seen from the distance of half a century it is tempting to portray this period as a 'golden age', free from the difficulties and frustrations normally accompanying scientific research. This would be quite wrong, and difficulties there undoubtedly were, though time may have smoothed the edges of these in the recollections of those involved. Primitive equipment and lack of resources were a problem for some groups (notably the Paris workers); difficulties in personal relations (e.g., Tjio and Levan) and difficulties in recognition and respect for women as scientists (as mentioned above) were all present at some point; while competitiveness between groups was certainly present, though not the secretiveness and aggressive competition that characterised some of the early work in molecular biology. All in all, those involved in the beginnings of human cytogenetics give the impression of being happy as a community, as well as individuals.

Turning to more general themes in the development of science, there are several that stand out and which have already been flagged in earlier chapters. Most striking is the power of the already accepted idea, in giving a preconception that can profoundly modify subsequent results and conclusions. I know of no other example, at least in the biological sciences, that is as striking as that of the human chromosome number, where acceptance of the perceived number of 48 was successively confirmed over a period of more than 30 years by numerous investigators, before the true number of 46 was finally determined. This persistent misinterpretation is all the more remarkable since none of those involved had any personal advantage to be gained from the erroneous finding, and all firmly believed that they were being objective in the interpretation of their work. The earlier workers (e.g., Winiwarter) stressed the approximate nature of their findings due to limitations of technology, yet even those with all the new techniques at their disposal 30 years later, were still able to misinterpret their results, as seen in TC Hsu's erroneous result and subsequent lament quoted in Chapter 2.

This episode should serve as a powerful reminder that very few scientific results are as objective as they seem, and that our interpretation of raw data is powerfully influenced by the general and personal framework that is a necessary part of any meaningful interpretation. An echo of this situation can be seen also in the opposition that Paul Polani's interpretation of the male sex chromatin pattern in Turner syndrome encountered; in this case the new conclusion had to overcome 50 years of thinking on the chromosomal basis of sex determination, conditioned by the original work on *Drosophila* and other insects.

More generally accepted in the history of science, but a striking example nonetheless, is the overwhelming importance of technology in permitting new discoveries. Despite the talent and devotion of the early cytogeneticists before 1950, the available techniques were simply not adequate to study mammalian chromosomes in detail. Likewise, it was not just a single technical advance that was needed to transform the situation but the application of a series of new techniques simultaneously.

Rather more puzzling is how some of these technical advances had to be discovered not just independently, but repeatedly. The use of hypotonic solutions in chromosome spreading must surely hold the record here, with at least five apparently independent discoveries over a 30-year period – Andres, Hughes, Hsu, Makino, and finally Nowell in 1960. Nor is this the only example, the use of mitotic stimulants allowing culture of normal peripheral blood having again been described in detail (and in a British journal) by the Russian workers almost 30 years before Nowell's use of phytohemagglutinin. However closely communicating the small human cytogenetics community may have been in the late 1950s, it is hard to disagree with Kottler that they were at times lacking in their awareness of their predecessors' work.

One final feature of human cytogenetics that does not feature in this book, but which has impressed me strongly, is its persistence and durability, in large measure due to the successive advances in technology continuing after 1960, and outlined all too briefly in Chapter 8. Since molecular biology began to have applications in human genetics and medicine towards the end of the 1970s, its practitioners have confidently and regularly predicted the demise of human cytogenetics as a useful field of science. Yet its resilience and, more recently its adoption of molecular techniques, have meant that, 50 years after its beginnings, scientists still study human chromosomes under the microscope, that original discoveries still result from their work, and that clinicians and their patients still benefit from this knowledge. It is a tribute to the work and workers described in this book that the foundations which they laid half a century ago should have begun the transformation of our knowledge not just of human chromosomes, but of human genetics generally, in a way that continues to have profound effects today.

APPENDIX 1

Some general historical publications on human cytogenetics

Capanna E (2000). Chromosomes yesterday: a century of chromosome studies. *Chromosomes Today* **13**, 3–22.

Carlson EA (2004). *Mendel's Legacy. The Origin of Classical Genetics*. Cold Spring Harbor, CSHL Press.

Chu EHY (2004). Early days of mammalian somatic cell genetics: the beginnings of experimental mutagenesis. *Mutation Res.* **566**, 1–8.

Ferguson-Smith MA (1993). From chromosome number to chromosome map: the contribution of human cytogenetics to genome mapping. *Chromosomes Today* **11**, 3–19.

Hamerton JL (1971). *Human Cytogenetics*, Vol. 1, Chapters 1 and 2. New York, Academic Press.

Harnden DG (1996). Early studies on human chromosomes. *BioEssays* **18**, 163–168.

Harper PS (ed.) (2004). *Landmarks in Medical Genetics. Classic Papers with Commentaries*. New York, Oxford University Press.

Harris H (1995). *The Cells of the Body. A History of Somatic Cell Genetics*. Cold Spring harbor, CSHL Press.

Hsu TC (1979). *Human and Mammalian Cytogenetics. An Historical Perspective*. New York, Springer.

Hungerford DA (1978). Some early studies on human cytogenetics. *Cytogenet. Cell Genet.* **20**, 1–6.

Lima de Faria A (2003). *One Hundred Years of Chromosome Research*. Dordrecht, Kluwer.

Makino S (1975). *Human Chromosomes*, Chapter 1. Tokyo, Igaku Shoin.

McKusick VA (2001). A history of medical genetics. In: *Emery and Rimoin's Principles and Practice of Medical Genetics*. London, Churchill-Livingstone, pp. 3–6.

Penrose LS (1966). Human chromosomes, normal and abnormal. *Proc. Roy. Soc. London B* **164**, 311–319.

APPENDIX 2

List of those interviewed

The following individuals were interviewed in relation to early studies on human cytogenetics. In most cases the interview was recorded and a transcript made, which is available on a confidential basis for those undertaking historical research. Material from some of the interviews appears (with permission) as quotations throughout this book, while excerpts from some of the recordings are given on the accompanying CD (see Appendix 3).

1. Ewart (Mike) Bertram (jointly with Keith Moore), Toronto**
2. Joy Delhanty, London*
3. John Edwards, Oxford**
4. Edward Evans, Oxford**
5. John Evans, Edinburgh*
6. Malcolm Ferguson-Smith, Cambridge**
7. Marco Fraccaro, Pavia*
8. Marthe Gautier, Paris*
9. John Hamerton, Winnipeg**
10. David Harnden, Manchester**
11. Kurt Hirschhorn, New York*
12. Maj Hultén, Birmingham**
13. Patricia Jacobs, Salisbury**
14. Muriel Lee, Edinburgh**
15. Jan Lindsten, Stockholm*
16. Mary Lyon, Harwell*
17. Eva and Yngve Melander, Lund
18. Ursula Mittwoch, London*
19. Keith Moore, Toronto (jointly with Mike Bertram)**
20. Peter Nowell, Philadelphia (by correspondence)
21. Paul Polani, London**
22. Lore Zech, Uppsala*

*Indicates interview recorded, but not on the CD; **indicates excerpt of recording on the CD.

APPENDIX 3

The recordings: key to excerpts included on the CD

THE RECORDINGS CONTAINED IN THE attached compact disc are excerpts from longer recordings forming part of a series of interviews with early workers in human and medical genetics between 2003 and 2005. They are given here with the consent of those involved; transcripts of the full interviews are available for consultation.

It should be noted that the primary purpose of the interviews was to provide a written record, not a sound archive, of the memories and experiences of the workers concerned. Recording quality is thus variable and no attempt has been made to control background noise. The excerpts are included here since many people suggested that readers would appreciate hearing the voices of those involved with some of the discoveries, and a description of their work in their own words.

1. Introduction
2. Professors Ewart (Mike) Bertram and Keith Moore
3. Professor Maj Hultén
4. Professor John Hamerton
5. Dr Edward Evans
6. Professor Paul Polani
7. Professor Malcolm Ferguson-Smith
8. Professor David Harnden
9. Professor Patricia Jacobs
10. Mrs Muriel Lee
11. Professor John Edwards

Index

Aborigine, Australian, 35
Air drying, 141
Allium, 6, 31
Andres, AH, 16–17, 118, 171
Anencephaly, 101
Ascaris, 3
Autolysing red blood cells, 17, 139–140
Autoradiography, 163
Autosomal trisomies (13, 18), 97–108
Awa, A, 117

Balbiani, EG, 3
Banding,
    centromeric (C), 161
    chromosome, 22, 118, 159–162
    reverse (R), 161
Barr, Murray, 2, 18, **19**–22, 78
Bateson, William, 4
Belling, J, 11
Bertram, Ewart, 2, 18, **19**–22, 78, 167
Biesele, J, 31
Biopsies,
    bone marrow, 55, 85, 97, 119
    chorion villus, 162
    skin, 56, 70, 101, 138
    testicular, 55, 84
Blakeslee, AF, 108
Blood culture, 17, 118–121, 123, 139–141
Böök, Jan, 58
Boveri, Theodor, 2, **4**, 126
Buccal smear, 21, 78
Burkitt lymphoma, 161

Callan, M, 87
Camera lucida drawings, 11, 16, 146
Cameron, Hugh, 102
Cancer,
    chromosomes in, 4, 31, 117–118
    genetics, 161

Carlson, EA, 5, 77
Carr, David, 108, 158
Carter, Cedric, 69
Caspersson, Torbjörn, **160**
Catcheside, David, 143–144
Cats, 20–21
Cell nucleus, 3
Chromosome,
    5p- syndrome, 157
    9q, 125
    22, 123, 161
    anomalies, 97, 156
    arms, 148
    groups, 147
    identification, 161
    morphology, 163
    paints, 164
    pairing, X–Y, 159
    Philadelphia, 125, 161
    syndromes, 155–158
    X, 21, 77–90, 159
    X inactivation, 90, 159
    Y, 9, 78, 89–90, 123–124, 162
Chromosomes,
    and heredity, 4
    as cell structures, 2
    cancer, 4, 31, 117–118
    diploid number, 9
    in Down's syndrome, 55–75
    early studies of human, 5–11
    and gene mapping, 162–163
    human number, 6, 7, 29–40, 171
    insect, 5, 77
    in leukaemia, 98, 117–126
    radiation-induced damage, 82
    sex, 77–90, 146
    spontaneous abortion, 158
Chrustchoff, GK, 17, 139
Chu, EH, 117, 138, 145

Classification, 142–149
Cleft palate, 106
Colchicine, 12, 31
Collaboration, between scientists and clinicians, 169
Colour blindness, 81
Comptes Rendus, 62, 63
Congenital abnormalities, 97, 99–108, 156
Congenital heart disease, 78–79
Court Brown, Michael, 84, **119**–120, 125, 168
Cultured cells, 12, 137–138
Cytogenetics (*see also* Chromosome(s)),
  cancer, 31, 117–118
  clinical, 71, 155–164
  diagnostic, 137, 162
  plant, 11, 30–31
  population, 89, 156
  studies in leukaemia, 17, 118–126

Dahlberg, Gunnar, 65
Darlington, Cyril, 11
Darwin, Charles, 1
Datura, 101, 104, 107–108
Davenport, Charles, 56
de Grouchy, Jean, **157**
de la Chapelle, Albert, 90
Deletions, 156
Delhanty, Joy, 67, 69, 158
Denver Conference, 98, 103, 142–149, 170
Dermatoglyphics, *see* Fingerprint patterns
DNA, 21, 163, 164
  sequencing, 163
Down's Syndrome, 55–71, 101, 123
Drosophila melanogaster, 5, 77, 80–81, 87, 171
  giant salivary gland chromosomes, 5, 8
Duchenne muscular dystrophy, 163
Dysmorphology, 103

Edwards, John, 70, **99**–100, 138
Edwards Syndrome, *see* Trisomy 18
Evans, Edward 33
Evans, HM, 7
Evolution, 1

Ferguson-Smith, Malcolm, 66, **83**, 146, **157**, 164
Fibroblasts, 59
  skin cultures, 35, 56, 70, 97, 123, 137–138
Fingerprint patterns, 58

Flemming, Walter, 3
Ford, Charles, **32**, 58, 66, 69, 81–82, 84–85, 101, 119, 142–143
Fraccaro, Marco, 58, 65, **66**
Funding bodies, flexibility, 169–170

Galton, Francis, 2
Galton Laboratory, 57, 58, 65, 69
Gautier, Marthe, 58, 59, **61**–64, 138, 169
Genetic linkage analysis, 163
German, James, 158
Gilgenkrantz, Simone, 59
Glass, Bentley, 8
Grubb, Rune, 31, 37
Gustavsson, Å, 36

Haldane, JBS, 139–140
Hamerton, John, 32, **33**, 69, 147
Harnden, David, 35, 58, 66–67, **100**–101, 138, 145, 168
Harris, Henry, 2, 168
Hauschka, T, 31, 117
Henking, H, 77
Heterogametic, 77, 81
Hirschhorn, K, 65, 90, **157**
Hofmeister, W, 2
Homogametic, 77
Hsu, TC, 1, 8, **13**, 15–16, 40, 55, 117, 145, 168, 171
Hughes, A, 15, 171
Hultén, Maj, 35, 168
Human embryonic material, 11, 16, 36
Human Genome Project, 163
Hungerford, David, 120-**122**, 123–125, 140, 157, 168, 170
Hydatidiform moles, 101
Hypotonic solutions and chromosome spreading, 13, 15, 117, 119, 121, 122, 171

Inheritance,
  early concepts, 1
  mechanism, 2
In situ hybridisation, 163
  fluorescent (FISH), 164
International System for Human Cytogenetic Nomenclature (ISCN), 148, 162
International Human Genetics Congress, First, 33, 59
Interviews, 175

Jacobs, Patricia, 64–65, 84-**87**, 88–90, 124, 168

Jost experiments, 80

Kemp, Tage, 12, 16
Klinefelter Syndrome, 22, 55, 78, 82–89, 156
Klinger, Harold, 148, 158
Koller, P, 117
Kottler, Malcolm, 9, 11, 39–40, 172
Kullander, Stig, 37

Lajtha, Lazlo, 85, 119
Lamy, Maurice, 64, 157
Lancet, 69, 79, 81–82, 98–99
Lee, Muriel, 87–88, 167
Lejeune, Jérôme, 58, 59, **61**–64, 69, 157, 169
Lennox, B, 83
Leukaemia,
  acute, 120
  chronic myeloid, 118, 120, 123–125, 161
Levan, Albert, 2, 12, 29, **31**, 36–40, 117, 138, 167
Levit, S, 17
Liège, 6, 8
Lindsten, Jan, 58, 65, **67**, 90
Loutit, J, 119
Lund, 12, 29, 30, 35–38
Lyon, Mary, **159**

Makino, Sajiro, 8, 10, **15**, 117, 121, 171
Malformation syndromes, 97, 99–106, 108, 161
Matthey, R, 13, 16, 21
McClung, CE, 77
Meiosis, 3, 9, 33, 40
Melander, Eva and Yngve, 37–39
Mendel, Gregor, 1
Miller, OJ, 66–67, 158
Microdeletion syndromes, 161
Mitosis, 3, 117, 141
Mittwoch, Ursula, 57, 68–69, 77, 159
Molecular genetics, 64, 163–164
Molecular techniques, 162, 172
Mongolism, *see* Down's Syndrome
Moore, Keith, 19–20, 78, 167
Morgan, TH, 5
Mosaicism, 70–71, 84, 108, 138
Muller, HJ, 143–144
Munzing, A, 36

Nobel prize, 64, 67
Nomenclature, 67, 142–149
  Denver, 125, 142–146
  Paris, 68, 142–143
Nowell, Peter, 120, **121**–125, 140, 170–171
Nucleolar satellite, 20

Oguma, K, 6, 7
Opitz, John, 104–105

Painter, Theophilus, 2, **8**–9, 39–40, 78
Pangenesis, 2
Patau, Klaus, 103–**104**, 142, 147, 157
Patau Syndrome, *see* Trisomy 13
Paulmier, F, 77
Penrose, Lionel, 17, 57, **58**, 65–69, 81, 158
Peripheral blood, culture, 123, 138–141, 171
Photomicrography, 11, 16, 32
Phytohemagglutinin (PHA), 16, 121, 140–141, 172
Polani, Paul, 58, 66, 68–69, 78–**79**, 80–82, 168, 171
Polydactyly, 106
Pomerat, CM, 13
Preconceived ideas, 16, 57, 171
Premature ovarian failure, 89
Prenatal diagnosis, 64, 162
Puck, Theodore, 12, 117, 138, 142–145

Quinacrine mustard, 161

Radiation-induced leukaemia, 85
Remak, R, 2, 3
Retinoblastoma, 161, 163
Rowley, Janet, 161

Schleicher, W, 3
Seabright, Marina, 161
Sex,
  chromatin, 18, 19, 21–22, 55, 78, 80, 83, 86, 89, 98, 159, 162
  determination, 5, 77, 80–81
  -reversed XX male, 84
  vesicles, 83
Skin fibroblast cultures, 35, 56, 137–138
Sloan–Kettering Institute, 37, 117
Smith, David, 103–**106**
Somatic cell hybrids, 163
Spermatocytes, haploid chromosome number, 7, 10
Spermatogenesis, in Klinefelter syndrome 83
Spermatogonia, diploid number, 7, 10, 33
Spontaneous abortion, 108, 158, 161
Squash technique, 11–12, 16

Staining,
  fluorescent, 160–162
  Giemsa, 161
Stalin, 17
Stern, Curt, 143–144
Stevens, Nettie, 77
Stich, Hans, 12
Strong, John, 64, 84–85
Sutton, Walter, 2, 5, 77

Texas, University of, 8, 39
Therman, Eeva, 103–**105**, 157
Tissue culture methods, 12, 117
Tjio, Joe Hin, 2, 12, 29, **30**–32, 36–40, 117, 142–143
Translocation, 63
  Down's Syndrome, 63, 68–70, 98
  reciprocal, 161
Triple X, 89, 98
Triploidy, 57, 98, 108, 158
Trisomy,
  13 (D), 97, 98, 103–105
  18 (E), 97, 98, 100–103
  21, 59–63, 68, 98, 103, 123–125, 146, 158
  in *Datura*, 101, 107–108
Turleau, Catherine, 157
Turner Syndrome, 22, 55, 68, 78–82, 98, 158, 171

Turpin, Raymond, 57, 58–59, **62**, 157, 169

van Beneden, E, 3
Vicia faba, 6
Virchow, R, 3

Waardenburg, P, 56
Weissmann, A, 2
Westergaard, M, 68
Wilson, EB, 4, 77
Winiwarter, Hans, 2, **6**, 7–10, 168, 171
Wolf, U, 157
Women as scientists in cytogenetics, 168–169
World Health Organisation, 68

X chromosome,
  biology, 159
  inactivation, 159
XX males, 159
XXX females, 89, 98
XXYY Syndrome, 8
XY bivalents, 40, 83
XYY Syndrome, 89–90, 156

Zech, Lore, 160–161
Zhivago, PI, 17